ALSO BY JAMES TABERY

Beyond Versus: The Struggle to Understand the Interaction of Nature and Nurture

TYRANNY of the GENE

TYRANNY of the GENE

PERSONALIZED MEDICINE AND ITS THREAT TO PUBLIC HEALTH

James Tabery

ALFRED A. KNOPF NEW YORK 2023

THIS IS A BORZOI BOOK
PUBLISHED BY ALFRED A. KNOPF

www.aaknopf.com

Knopf, Borzoi Books, and the colophon are registered trademarks of Penguin Random House LLC.

Grateful acknowledgment is made to Sarah Keim for permission to reprint emails dated September 25, 2006 and May 17, 2007; and Peter Scheidt for an email dated January 14, 2004.

Library of Congress Cataloging-in-Publication Data
Names: Tabery, James, [date]- author.
Title: Tyranny of the gene : personalized medicine and its threat to public health / James Tabery.
Description: First edition. | New York : Alfred A. Knopf, 2023. | Includes bibliographical references and index.
Identifiers: LCCN 2022031887 (print) | LCCN 2022031888 (ebook) |
ISBN 9780525658207 (hardcover) | ISBN 9780525658214 (ebook)
Subjects: MESH: Genomic Medicine—legislation & jurisprudence | Genomics—legislation & jurisprudence | Precision Medicine | Environmental Health—legislation & jurisprudence | Politics | United States
Classification: LCC QH447 (print) | LCC QH447 (ebook) | NLM WB 33 AA1 |
DDC 572.8/6—dc23/eng/20221026
LC record available at https://lccn.loc.gov/2022031887
LC ebook record available at https://lccn.loc.gov/2022031888

Jacket illustration by Justin Metz
Jacket design by Tyler Comrie

Manufactured in the United States of America
First Edition

For Dawn-Marie

Contents

TYRANNY of the GENE

Introduction

ON AUGUST 16, 2011, my father, Mike Tabery, woke up and couldn't get out of bed. Until then, he was an avid outdoorsman. He tied his own flies, casting for rainbow trout on the upper Delaware River each spring, and then in the fall took out the German wirehaired pointers that he raised from pups to track, point, and flush grouse and pheasant. Saturdays in the summer were spent astride his riding lawn mower on a one-acre plot of land in the Pocono Mountains of Pennsylvania, a can of Pabst Blue Ribbon in his hand. Now sixty-four and recently retired, he anticipated a relaxing future doing the things he loved.

That Tuesday in August, however, my dad's legs wouldn't listen to what his brain was telling them to do. After he took an ambulance to the hospital and spent hours in an emergency room confused about what could possibly explain a sudden onset of paralysis, the diagnosis shook our family: stage 4 non-small-cell lung cancer. Non-small-cell lung cancer is the most common form of that disease, accounting for more than 80 percent of the cases. Beyond his lungs, there were tumors in my father's brain, an arm, ribs, pelvis, and spine. The tumors on his vertebrae quietly grew until they compressed his spinal cord, cutting off nerve signals to the lower portion of his body.

Lung cancer often strikes with merciless surprise. It is one of the most common cancers in the United States and in the world. Well over 200,000 Americans are diagnosed with it every year; that translates into a new diagnosis roughly every two to three minutes. Almost one-quarter of all cancer deaths in the United States start as a tumor in a lung.

Lung cancer is so deadly because the disease typically doesn't reveal itself until after it has spread to other parts of the body. Lungs have few pain receptors, so a person with a tumor in one or both of their lungs doesn't sense the growth, and they can't reach in and feel a lump. Eventually, a cancerous cell sloughs off and travels via lymph or blood to another part of the body. When it binds to healthy tissue in one of those new spots, another tumor can begin to grow, and symptoms such as tingling in the toes, memory loss, or pain around the ribs and back appear. It's at this point that the person seeks medical attention, but now that the cancer has metastasized, treatment options are limited.

My father was soon transferred from the emergency department to an oncology floor. There, the team of health-care providers laid out a coordinated plan for combating the advanced disease. He received radiation on the tumor sites and surgeries to stabilize several bones made dangerously fragile by the cancer; painful biopsies allowed physicians to examine the cancerous tissue itself. He was often in an operating room or an MRI machine and had physical therapy sessions designed to keep his muscles limber in the hope of regaining mobility.

The metastatic cancer turned my mother's world upside down as well. The hospital where Dad received care was about an hour away from their rural home, so each day she made the long, often lonely commute to be with him during those chaotic early weeks. Entries on her calendar for "babysit granddaughter" and "lunch with Dan and Darryl" got crossed out and replaced with "spinal radiation" and "MRI-brain."

After what felt like an eternity of bad news, our family received some hope when the results of a test revealed that the cancerous cells in my dad's body had a molecular marker on their surface making them an ideal candidate for a drug that could target those cells and disrupt their growth. Erlotinib (brand name Tarceva) is a textbook example of what is often called "personalized medicine" or "precision medicine." (The relationship between personalized medicine and precision medicine is actually a point of contention, and just how this came to be the case will become clear later. For now, we can just treat them as synonyms.) The personalized medicine motto is "the right treatment, for the right patient, at the right time." The goal is to tailor health care to the individual characteristics of patients, in contrast to traditional one-size-fits-all medicine. On the traditional model, the criticism goes, all of us are treated as if we were identical humans. Personalized medicine, instead, aims to take the differences between us into account when designing precise plans for medical care. Erlotinib, for example, was initially approved by the U.S. Food and Drug Administration (FDA) in 2004 to treat any patient with non-small-cell lung cancer. But it soon became clear that the drug didn't work well for everyone. A study published just months before my dad's diagnosis showed that erlotinib was most effective for people with his specific genetic mutation. Just weeks after he took his first dose, my mother phoned with tears of joy in her voice to tell me that Dad's tumors were shrinking.

Champions of precision medicine have hailed its move away from the traditional model of health care as "a novel paradigm," "revolutionary," and "a new era" in medicine. The precision approach has gained its greatest traction in the treatment of cancers. It also has been deployed to combat rare conditions like sickle-cell disease. The ultimate vision, though, is for this revolution to spread across all aspects of health care. Efforts are under way to apply the model to psychiatry, dentistry, gynecology, dermatology, physical therapy, ophthalmology, public health, and spine surgery, to name just a few fields. At this point, there are very few health sciences where evidence of this precision approach can't be found.

The success of drugs like erlotinib has generated inspiring claims about the future of health care driven by this approach. "Miracle cure," "magic bullet," "biomedical game changer," "breakthrough drug"—these are just some of the phrases that orbit precision medicine. Enthusiasts foresee a fundamental transformation of the medical landscape. As health-care costs continue to skyrocket, precision medicine proponents envision a more affordable alternative, one that tailors interventions for patients from the beginning of treatment rather than sending them off on a trial-and-error odyssey, thereby decreasing waste while increasing quality of care. Moreover, the promise is that precision medicine research will ensure these cures and breakthroughs extend to all communities, thereby combating unjust racial and socioeconomic health disparities. With such scientific potential, such widespread applicability, such cost-effectiveness, and such aspirational inclusivity, it's hard not to get excited about a future guided by this medical revolution.

A closer look at lung cancer generally, however, and my father's experience with it specifically, paints a far gloomier picture than what the personalized medicine advocates would have you believe. Many of the claims surrounding personalized and precision medicine, as we'll see, are wild and dangerous exaggerations. The promises are genuinely fueling a fundamental transformation in medicine, but it's a revolution that should make all of us profoundly concerned about what the future holds for our health and our health care.

. . .

The months following my dad's diagnosis were disorienting for our family. The plan was to transition him from the hospital to home by way of a stopover in a rehabilitation facility where he'd get help regaining his mobility, but he injured his knee in rehab and had to return to the hospital for surgery. Then back to rehab. Days afterward, a routine chest X-ray revealed blood clots in his lungs that could have proven deadly, so he underwent a surgery during which a filter was placed in one of his leg veins. When he returned to the rehabilitation center for the third time, he acquired a staph infection that spread across his body and defied standard antibiotics. An emergency consult with an infectious disease expert followed.

By that point, the bills for the care he'd received since August started arriving. We soon were calling Dad the "million-dollar man," and it wasn't an exaggeration. Surgeries, radiation, ambulance rides to the emergency room, treatments in the emergency room, MRI scans, blood tests, X-rays, days in the hospital, weeks in rehab—the cost was impressive. The real eyepopper, though, was the erlotinib: a one-month supply was billed at more than $5,000.

Personalized medicine enthusiasts promote it based in part on the idea that it will lead to dramatic cost savings for the health-care industry and for patients and their families. But that vision is severely compromised by the fact that personalized medicines themselves are exorbitantly expensive. Erlotinib is actually on the affordable end of the spectrum, where you can also find drugs with list prices in the $50,000-per-month range. Precision medicine drugs, in fact, are among the most expensive pharmaceuticals in existence, with some priced in the millions of dollars for onetime treatments. There is a straightforward biological reason for this. Personalized medicine works by carving up patient populations into smaller and smaller groups based on their specific molecular-genetic profiles. For example, a population like "patients with lung cancer" is divided into "patients with lung cancer who have the epidermal growth factor receptor mutation" (like my dad) versus "patients with lung cancer who do not have the epidermal growth factor receptor mutation." There are about a dozen such biological distinctions made now with non-small-cell lung cancer. That creates smaller markets for the pharmaceutical companies that research, develop, and manufacture the drugs—companies that are, first and foremost, in the business of making a profit. To offset the relatively small number of consumers who are eligible for their new drugs, pharmaceutical companies drastically hike up the prices. Personalized medicine purports to both individualize care and drive down health-care costs, but the more it succeeds at individualization, the higher the prices.

In a recent study that tracked non-small-cell lung cancer patients between 2000 and 2011, as precision medicines like erlotinib entered the medical marketplace and started replacing standard chemotherapies designed for broad use, researchers found that patient spending on their health care increased. Spending on inpatient treatments went down because less time was spent in the hospital, but average costs went up because of the expensive treatments. Higher out-of-pocket expenses for drugs like erlotinib, another investigation revealed, translated into decreased rates of survival because some patients couldn't consistently afford the pricey pills. In fact, oncologists have begun warning their patients of a new side effect of cancer care in the age of precision medicine—financial toxicity. People newly diagnosed with cancer are at increased risk of using up their life's savings, losing their home, and going bankrupt.

Our family was fortunate. My father's medical care was covered by my mother's employer-based health insurance. Out-of-pocket expenses added up to tens of thousands of dollars, but my parents were fairly secure financially, and they had an established network of friends and family who could offer nonfinancial aid. In November, my father came home. He still couldn't walk or even stand, but he could wiggle his toes. We welcomed that progress. My mother took on the Herculean task of being his primary caregiver—coordinating physician appointments, managing two dozen prescriptions, helping him to continue physical therapy, making meals, tending wounds. Friends kindly built a wheelchair-accessible ramp so that Dad could enter and leave the house. A co-worker let them borrow a wheelchair-accessible van for the scheduled trips to and from the hospital. Others helped transform the living room of my parents' home so that it was still cozy enough for friends and family to spend time chatting on the couch or catching a Phillies game on television but also clinical enough so that Dad could get the care he needed there. An elaborate lift system was constructed to hoist and move all 250 pounds of him from the bed to his La-Z-Boy, where he continued tying colorful, meticulous flies in anticipation of the day when he could wade back out into the chilly spring waters of the Delaware.

Dad had another thing going for him: race. Physicians have long known that Black and white patients move through the process of being treated for lung cancer differently, with sometimes deadly consequences. The best chance for surviving lung cancer has always been catching it early and surgically removing the tumor before it metastasizes. Those surgeries, though, are performed less often on Black patients than on white, and that in turn means lower survival rates for Black patients. A 2019 study revealed that white patients with lung cancer were almost twice as likely as Black patients to get access to the diagnostic tests necessary to see whether a drug like erlotinib was a good match for them. Patients from richer neighborhoods were also more likely to be tested than patients from poorer neighborhoods, regardless of race. The arrival of drugs like erlotinib, unfortunately, hasn't combated the racial and socioeconomic health disparities associated with lung cancer; it has just created a new node of inequities in the health-care process.

A period of joy and stability set in after my dad returned home.

Our whole family—Mom, Dad, my wife and I, my brother and his family—spent Thanksgiving and then Christmas together. Aunts and uncles, nieces, in-laws, neighbors, co-workers, and friends all visited regularly. My parents were rarely alone in their house throughout the first half of 2012. In June, we spent Father's Day together, and Dad turned the tables by surprising everyone in the family with matching T-shirts advertising "Tabery's Bait & Tackle," a reference to some fictitious fishing shop that sold make-believe lures and bobbers.

In July, however, my parents went to the hospital for a regular checkup, and a scan revealed that the tumors had begun growing again. The cancerous cells had developed a resistance to the drug. Erlotinib and drugs like it just put cancer on pause, and when it comes back, it comes back with a vengeance. My dad's body deteriorated rapidly before he could be shifted to a new course of treatment. He never regained the ability to stand on his own, let alone walk or cast a fly on the Delaware. On September 23, thirteen months after he woke up paralyzed, my father died.

. . .

Advocates of personalized medicine forecast a future in health care revolutionized by the arrival of miracle cures and magic bullets. In reality, personalized medicine's flagship domain of cancer care reveals a landscape where genuine cures are still few and far between. For patients like my dad with advanced cancers, the actual benefit typically means just months added to the end of life.

Champions of personalized and precision medicine go wrong in thinking that the best way to improve the health of patients and populations, to cut the rising costs of health care, and to narrow health inequities is to orient health care around the genetic differences between us because, as we'll soon see, "the right treatment, for the right patient, at the right time" has always been about prioritizing DNA to decide what's "right." In contrast, decades of scientific research have pointed to a much more reliable route to meeting those goals: making the physical, chemical, biological, and social environments in which we live safer and healthier. It's easy to take for granted how important it is to have access to a nutritious diet, how dangerous it is to come into sustained contact with pesticides, how hazardous it is to expose children to lead, and how essential it is to lay a baby to sleep on her back. But those realizations were all once points of contention which were settled by public health research that attended to the impact of the environment on our health.

Lung cancer is a prime example. We now know it is caused primarily by exposure to substances such as tobacco smoke, asbestos, and radon and that the synergy between them can be particularly dangerous. We'll never know for certain what caused my dad's lung cancer, but all three of those risk factors figured into his life at one point or another. Radon is measured in picocuries per liter, and the U.S. Environmental Protection Agency (EPA) encourages homeowners to keep their indoor radon levels below 2; anything above 4 is real cause for concern. In the Columbus, Ohio, neighborhood where my father grew up in the 1950s, average indoor radon levels are more than 8. He was also exposed to a steady stream of cherry Cavendish tobacco smoke from my grandfather's pipe in those childhood years. When he grew older, secondhand smoke became firsthand smoke. By the time my dad was diagnosed with lung cancer, he hadn't touched a tobacco product in three decades, but he smoked cigars and cigarettes heavily in his twenties. Most patients with non-small-cell lung cancer aren't active smokers or never smokers; they're former smokers like him. In the late 1970s and early 1980s, just as I was born, my dad was between jobs and brought in money for our family as a construction laborer, where he very likely came into contact with asbestos products. Soon after, when my dad found steady employment working at the county jail, we moved to the small town of Saylorsburg, Pennsylvania, where indoor radon levels dangerously average above 15 picocuries per liter. The region sits adjacent to the Reading Prong, a uranium-rich vein of metamorphic rock that generates some of the highest radon levels in America.

Benjamin Franklin famously advised, "An ounce of prevention is worth a pound of cure." He was warning Philadelphians about fire prevention, but the wisdom applies just as well to health concerns. For most diseases, taking steps to prevent the development of illness is both more effective and more cost-efficient than trying to cure the disease after it arrives. There is still no cure for metastatic lung cancer, despite the steady arrival of expensive pharmaceuticals like erlotinib introduced to treat patients after they are diagnosed. The good news, though, is that rates of lung cancer in the United States have been going down steadily since the 1990s due to public health efforts aimed at making our environments more lung-friendly with radon remediation systems, asbestos abatement programs, and antismoking campaigns. Proponents of personalized and precision medicine hail its departure from traditional, one-size-fits-all medicine. But most harmful things in our environments—among them, automobile exhaust, factory smog, landfill runoff, tobacco smoke, industrial waste, contaminated drinking water, heavy metals, highly processed foods, work-related stress, economic insecurity, and infectious diseases—really are bad for everyone. That's why the health and economic impact of addressing them is so great: everyone benefits, not just the folks with the right genes, the right color skin, or the right bank account balance.

Communities of color and those living in poverty are particularly susceptible to the harm of unhealthy environments. They tend to be the ones who are closest to the factories, landfills, and highways. They are less likely to have reliable access to healthy foods, green spaces, and walkable neighborhoods. Health disparities, it is abundantly clear, are caused by differences in our environments, not differences in our genes. Black patients with lung cancer aren't dying at higher rates than white patients because their DNA makes them particularly susceptible to the deadly disease; they're dying at higher rates because they live in a world that places Black people in environments where the air is less safe to breathe and then treats Black people with lung cancer differently than white people with lung cancer.

The concerns I've raised here about personalized medicine—that much of the talk of "magic bullets" and "biomedical game changers" is more hype than reality, that precision medicine risks increasing rather than decreasing the costs of health care, that focusing on genes distracts from the actual causes of health disparities, that attending to the determinants of health in our physical and social environments offers a far more effective means of improving the health of patients and populations—aren't novel observations. In fact, these concerns about a tyranny of the gene were raised almost immediately after pharmaceutical executives coined and then started marketing "personalized medicine" as a new approach to health care in the late 1990s.

And yet, despite those warnings, the major players in American health research are embarking on a massive swerve down a path toward personalized medicine. Consider the prescription drug landscape. In 2000, you could count on one hand the number of pharmaceutical compounds that were designed for patients with certain molecular-genetic markers; in 2018, two out of every five new molecular drugs that the FDA approved were developed with the principles of personalized medicine. This transition is the result of a pharmaceutical industry that is shifting more and more of its research and development resources away from drugs that are designed to work for everybody and instead toward drugs that are designed to work for those patients who have the right biological makeup.

Hospitals are also driving this trend. The president and CEO of Geisinger—an enormous health system spread out over Pennsylvania and New Jersey that serves roughly three million patients—said in 2018 that he wanted to make obtaining the DNA of every Geisinger patient standard clinical care. To kick-start that effort, Geisinger launched its MyCode Community Health Initiative in collaboration with Regeneron Pharmaceuticals. The Geisinger patients are encouraged to sign up and provide their genetic information in exchange for the possibility that their doctors could learn something useful about their health. In reality, only about 5 percent of those who participate receive anything of clinical value. The genetic data, though, is handed to the drug company, which uses it to guide the development of new drugs and diagnostics from which both Regeneron and Geisinger profit. More such collaborations between hospitals and pharmaceutical companies have emerged—at Intermountain Healthcare in Utah, at the CU Anschutz Medical Campus in Colorado, at UCLA Health in California, and at Mount Sinai Health System in New York.

The federal government is encouraging this genetic swerve as well. In 2015, the National Institutes of Health (NIH) embarked on a historic endeavor, the All of Us Research Program, which aims to enroll a million or more Americans over the coming years and track their health for at least a decade. The primary design and purpose are to obtain DNA from the participants, thereby producing a vast data set that will permit studying how genetic differences between people translate into differences in health and treatment outcomes. Folks living in the Hill District of Pittsburgh and the Harlem neighborhood of Manhattan, both targeted because of their diverse populations, have already provided researchers with samples of their blood, from which their DNA has been extracted.

How did we get here? Why is biomedical research in America lurching toward placing the gene at the center of medicine even though there are now and have been for some time obvious reasons why we should be skeptical of the extent to which this genomic reorientation can actually deliver on its lofty promises? My aim here is to answer those questions by telling the story of where this zeal for personalized medicine came from; revealing who promoted it along the way and why they did so; assessing how it became so popular so quickly; considering who stands to benefit from it and who does not; identifying which alternate approaches to improving health it crowded out; and making clear what lies ahead if we stay on this path.

My thesis is that there have been and remain powerful financial, political, technological, and scientific forces that are driving this embrace of personalized medicine and promoting the idea of medicine as something genetic while simultaneously impeding the study of environmental determinants of wellness and disease. Genes have become far easier to study than environments. Probing DNA has benefited from technological developments that have eluded environmental health research, and medical genetic research has been subject to less partisan politicization than environmental health research. The result of all this is a biomedical research industry that is now prioritizing the study of genetic causes of health and illness not because those causes play a particularly large role in health outcomes but because those causes are faster, cheaper, more profitable, and more politically palatable than the environmental alternative. The products of this research industry will lay the foundation for the future of health and health care: which health-promoting policies get prioritized by your federal, state, and local governments; what health interventions being conceived today will be offered to you tomorrow; how your physician diagnoses, treats, and advises you; where we as a society place responsibility for being healthy and sick. We all have a stake in what this future holds.

. . .

The seeds of this genetic swerve were planted decades ago, in the mid- to late twentieth century; the movement, though, really took off after the completion of the Human Genome Project in 2003. Scientifically, the full sequence of the human genome laid the foundation for subsequent efforts to explore how differences between people's DNA coincided with differences in who got sick and who stayed healthy. There was also something profoundly symbolic about the Human Genome Project. It was characterized as providing humanity with the "book of life" and the "holy grail of biology." It allowed you to hold the full sequence of the human genome in your hand on a DVD, but the product of the effort was never meant to be an idle string of letters on a screen. As Francis Collins, director of the Human Genome Project, put it upon announcing the successful completion, the achievement "should not be viewed as an end in itself. Rather, it marks the start of an exciting new era—the era of the genome in medicine and health."

We must look first to the years preceding the Human Genome Project to see how figures, events, and ideas shaped what transpired in the twenty-first century—how a collection of dead songbirds helped inspire a book that changed the way we think about the environment, how a young man who barely escaped being sent to Auschwitz forever altered the way pharmaceutical companies research and develop their products, how the sudden death of President Franklin Delano Roosevelt reconfigured the way scientific research was conducted, how a brilliant innovation in sensor technology suddenly made accessing massive amounts of genetic data feasible, how the study of an Indigenous community in Arizona fueled the search for genetic causes of health disparities. That will set the scene for the story of what followed, the Human Genome Project, and the viral spread of "personalized medicine" as its supporters came into conflict with those determined to study the environment and its impact on public health.

Telling the story of the competing efforts to transform health care provides a lens through which we can see how the wider genetic swerve has unfolded. It lays bare the stark differences between the science of nature and the science of nurture. It makes clear the alternate visions for what the future of medicine is supposed to look like, one where it is tailored to DNA and one where it is oriented around safe environments. It reveals how social forces like private industry, technological developments, media narratives, and political pressures got behind and then amplified the promises of DNA-based personalized medicine. In short, this story lets us see how we reached the current situation where drug companies, hospitals, and the federal government are all reorienting health and health care around our genes, even though the value of the path ahead is more than slightly questionable.

. . .

Just how much time erlotinib added to the end of my father's life is as impossible to specify as what particular thing triggered his lung cancer. Still, there is every reason to believe that he lived longer because of the drug. I will always cherish that period with my dad. Not since I left home for college did our family spend so much time together. We grew closer, bonded by the exhausting and frightening experience of watching death steadily approach him. I still smile when I pull on my faded Tabery's Bait & Tackle T-shirt. Erlotinib, however, didn't cure him. For a very high price, the drug slowed his terminal disease down for a short period of time, and even that limited benefit is not available to everyone.

My son, Michael, never got to meet his namesake. He and his siblings were born years after their grandfather died. My hope is that they will grow up in a world that prioritizes making the places where they live, learn, play, and mature a safer space for all because it's the right thing to do, rather than in a world that prioritizes developing expensive treatments for the few because it's the lucrative thing to do. We can't turn toward that more equitable path, though, until we understand how we wound up on this one.

1

A Tale of Two Revolutions

OLGA OWENS HUCKINS and her husband, Stuart, had cultivated a two-acre oasis about thirty-five miles southeast of Boston on the Powder Point peninsula of Duxbury, Massachusetts. They'd left a large portion of it to wilderness, where reeds grew thick in freshwater ponds and mature cedars and oaks abounded. Just off the coast of Duxbury Bay, their refuge was a perfect spot for migrating birds to feed, rest, and nest. Olga and Stuart were bird lovers and welcomed birders onto their property to seek out the night herons and spotted sandpipers, the American goldfinches and Carolina wrens.

But on a summer day in 1957, Olga looked out at her beloved bird sanctuary in horror. The ground was filled with the corpses of birds, their beaks open but silent and their tiny claws scrunched up to their chests in anguish. Days earlier, a crop duster hired by the State of Massachusetts to spray a pesticide aimed at exterminating mosquitoes had crisscrossed the Huckinses' sanctuary. The chemical, DDT (dichlorodiphenyltrichloroethane), killed indiscriminately. The couple found seven songbirds dead almost immediately. The next day three more lay lifeless around their birdbath. The day after that they watched a robin drop off a branch. The bees and grasshoppers were also killed off. Ironically, it seemed as if only the mosquitoes survived.

When, in January 1958, Olga read in *The Boston Herald* assurances from a representative of the State of Massachusetts spraying program that the pesticide they were continuing to release was safe and effective, she wrote a letter to the paper describing the devastation to her property and the wildlife that lived on it, as well as the deep betrayal she felt. "They were birds that lived close to us, trusted us, and built their nests in our trees year after year." No one, she said, could witness the impact of that pesticide and deem it harmless to all but mosquitoes. Olga also sent a copy of her letter to a friend who had worked for the U.S. Fish and Wildlife Service. Was there anyone who could help halt the spraying of DDT?

That friend was Rachel Carson. By 1958, Carson was well aware of the threat posed by DDT. In fact, in 1945, when she was working for the Fish and Wildlife Service, she'd reached out to *Reader's Digest* to see if it would be interested in a story about the DDT testing that the service was conducting. DDT was hailed as a miracle of modern science, used widely by the American military during World War II, sprayed across islands in the Pacific to kill malaria-carrying mosquitoes, and doused on soldiers in Europe to kill typhus-carrying lice. After the war, DDT came to represent the Cold War mentality of humans triumphing over nature. In contrast to previous synthetic pesticides that tended to work only on certain pests, DDT promised to eradicate all sorts of vermin; what's more, the chemical compound was cheap and stayed in soil and on plants for extended periods, so it continued to repel insects long after the initial administration. When fire ants invaded farms across the South, the chemicals industry and the U.S. Department of Agriculture teamed up to respond with a campaign of widespread DDT spraying. Throughout the 1950s, crop dusters across the United States coated farms and marshes with a film of DDT, all in the name of boosting crop yield and exterminating pestilence.

In 1945, the *Reader's Digest* editors weren't interested in an essay from some unknown employee of the Fish and Wildlife Service. But by 1958 Carson had left the service and turned to working as a full-time writer, counting a trio of best-selling books about the wonders and ecology of oceans to her credit. She continued tracking the agricultural use and environmental impact of DDT, however, collecting technical reports and scientific publications on the topic. Carson began communicating with toxicologists, wildlife biologists, and chemists; she also joined forces with citizens like Huckins who were sounding the alarm about what pesticides were doing to their local communities. The research that Carson reviewed indicated that DDT accumulated in organisms that encountered it, and then grew more and more concentrated as it rose up the food chain. Entire populations of birds, fish, and small mammals were being wiped out by the pesticide. At the same time, insects repeatedly exposed to the chemical ended up gradually developing resistance, necessitating even more spraying.

In May 1958, just months after Huckins's letter reached her, Carson signed a book contract. Four years later, *Silent Spring* appeared. In it, Carson warned that pesticides like DDT were becoming deadlier and deadlier and ecosystems were being decimated and thrown out of balance. Carson didn't call for the complete ban of all synthetic pesticides, recognizing their value. But she did warn against the indiscriminate use of them given the dearth of information about the wider ecological and health impacts of sustained exposure to chemicals never before seen on Earth. *The New Yorker* serialized three chapters from Carson's book throughout June 1962, helping make *Silent Spring* an immediate sensation when it arrived in bookstores that September. Carson thanked Huckins for her 1958 letter in the acknowledgments, and the book's famed title was a nod to her as well, a reference to her songbirds' sad silence.

The chemicals industry, sensing a genuine threat, pounced. Because readers were given a peek into *Silent Spring* through *The New Yorker* excerpts, the chemical companies had prepared an all-out assault on Carson's book by the time it reached bookstores. Representatives from Monsanto Chemical Company and Dow Chemical Company accused Carson of being anti-science, anti-technology, and anti-progress. They warned of a planet overrun by insects and human populations threatened with starvation, epidemics, and misery. Carson's critics, almost entirely men, questioned her scientific credentials and routinely used gendered language to discredit her concerns. Her overtly "emotional" and "sentimental" diagnoses, they chided, were little more than "high-pitched" feminine hysteria and not at all conducive to the cold, hard fact-finding of proper science. Carson was limited in responding to her critics because she was battling breast cancer. She died in April 1964, less than two years after *Silent Spring* was published.

Carson's book and the widespread public interest in humanity's impact on the environment lived on. That attention and back-to-back ecological disasters in 1969 galvanized the modern American environmental movement. First was the Santa Barbara oil spill, when nearly 100,000 barrels of oil polluted the pristine Santa Barbara Channel off California and killed thousands of seabirds, dolphins, and sea lions. Just months later, a railcar passing through Cleveland threw a spark into the Cuyahoga River that landed on an oil slick, igniting the river in flames. Both of those events drew sustained national media attention, and it put pressure on politicians to act. Congressman Pete McCloskey of California and Senator Gaylord Nelson from Wisconsin helped organize the first Earth Day on April 22, 1970, and later that year President Richard Nixon ordered the creation of the Environmental Protection Agency. Congress, over the next three years, followed with the Clean Air Act, the Clean Water Act, and the Endangered Species Act. In 1972, the EPA banned the use of DDT in the United States.

Rachel Carson is remembered by history for launching the environmental movement, warning of the dangers to come if Earth isn't

Rachel Carson, interviewed in April 1963 on *CBS Reports,* warned of the dangers children faced from harmful chemicals.
CBS Photo Archive via Getty Images

protected from human activity. But *Silent Spring* also warned of the threats that humans posed to themselves. Pesticides like DDT, Carson explained, were in rivers and groundwater, in soil and plants, in the bodies of fish and domesticated animals, and therefore were "now stored in the bodies of the vast majority of human beings, regardless of age. They occur in the mother's milk, and they probably occur in the tissues of the unborn child." Once inside the body, those chemicals disrupted the immune system, altered metabolism, attacked the nervous system, and initiated cancerous cell growth. For millennia, public health was about sanitation and protecting communities from infectious diseases. Public health success stories involved keeping deadly microbes out of the human body, be it by separating sewage from drinking water or by quarantining the sick from the healthy. After *Silent Spring*, it became clear that public health needed to include protection from products that humans were producing to control nature.

The banning of DDT by no means solved the nation's environmental health problems. Chemical companies continued to produce new herbicides, fungicides, and insecticides. The exhaust from leaded gasoline pervaded the air that humans breathed. Toxic waste buried in the 1940s and 1950s was leaching up into playgrounds in the 1970s. Asbestos and lead paint surrounded people in the homes where they lived. These substances were associated with a range of alarming health outcomes, including miscarriages, birth defects, neurodevelopmental delays, respiratory disorders, and blood cancers.*

Research in the 1980s and 1990s made clear that the burdens of exposure to these harmful environments were distributed neither randomly nor equally. In 1982, Black Americans living in Warren County, North Carolina, protested when the state announced its intention to dump soil contaminated with toxic polychlorinated biphenyls in their community, one of the very few primarily Black

* The most famous episode from this period was the Love Canal disaster just outside Niagara Falls. The Hooker Chemical Company dumped twenty-one thousand tons of toxic waste in the canal in the 1940s and 1950s, buried it, and then sold the land to the Niagara Falls School Board, which built a school on top of the site; when the hazardous material leaked and percolated up to the surface in the 1970s, children were afflicted with epilepsy, birth defects, and asthma.

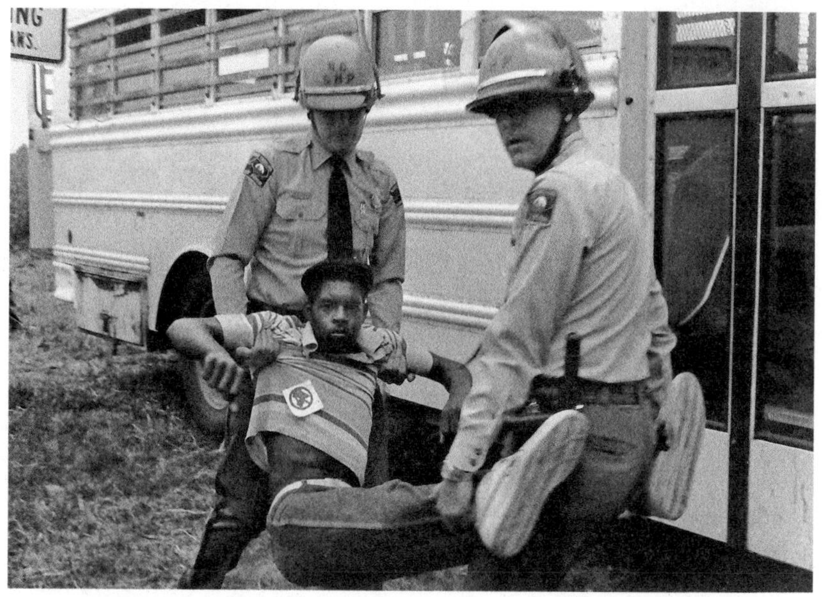

Black protesters, hundreds of whom were arrested, attempting to prevent the dumping, in 1982, of soil contaminated with polychlorinated biphenyls in their community in Warren County, North Carolina, sounding the alarm about what would soon be recognized as environmental racism.
Photographs by Jenny Labalme

counties in the state at the time. The water table in that rural region sat just ten feet belowground, so the residents reasonably feared that their supply of drinking water would be poisoned by carcinogenic runoff. Citizens, religious leaders, local politicians, and members of the NAACP fought back. Some lay down in the road, nonviolently preventing the trucks carrying the tainted soil from proceeding. Others linked arms and chanted, "They ain't gonna stop us now!" Marchers carried signs with "PCB" crossed out. Officers in riot helmets scooped up hundreds of protesters and carried them to a bus where they were transported to jail, and the state dumped the hazardous waste in Warren County just the same. But the massive civil disobedience drew national attention to a glaring prejudice. Subsequent research revealed that low-income communities of color with little in the way of political power were routinely prioritized for hosting industrial plants, oil refineries, factories, and landfills; the placement of those sources of pollution in the communities made the environments less healthy for the citizens and also less financially valuable, which in turn made the communities more likely to be candidates for hosting future toxic materials. The toxic feedback loop became recognized as a particularly harmful form of injustice, resulting from "environmental racism."

Around the same time, pediatricians started warning of the undue burden of toxic exposure felt by America's children. The EPA's system for testing and regulating safe levels of pesticides and other harmful chemicals was based on what the average person encountered and could tolerate. But pediatricians cautioned that exposure to even trace amounts of these substances could have vastly different impacts on the health of infants and young children, compared with adults, affecting their immune and endocrine systems, the development of their reproductive organs, and their neurocognitive function. Unfortunately, there was very little data on how sustained exposure to things like pesticides, noxious air, unclean water, and hazardous waste impacted children as they grew.

In 2000, politicians in Washington, D.C., decided it was time to gather that data. Congress passed the Children's Health Act, which called for a nationwide study of America's children that investigated the environmental impacts on their health. The study, by congressional mandate, had to be large and diverse enough to address racial health disparities; it had to last long enough that the children could be monitored until they were adults; and it had to be complex enough to consider all manner of environmental features—chemical, biological, physical, and psychosocial. On October 17, 2000, President Bill Clinton signed the Children's Health Act into law, officially calling for the National Children's Study.

. . .

In 1957, the same year that Olga Huckins discovered her dead songbirds in Massachusetts, a very different health revolution began taking shape on the other side of the United States when Arno Motulsky made the case for a new science of "genetically conditioned drug reactions." Motulsky, a medical geneticist at the University of Washington in Seattle, had taken a very circuitous route to reach that position. He fled Nazi Germany with his Jewish family in 1939 and made it to the Miami coast aboard the *St. Louis* with nine hundred other Jewish

Arno Motulsky (with arrow pointing at him) on the deck of the MS *St. Louis*.
AP Photo

refugees before the U.S. government denied the asylum seekers access to the country and forced the ship to return to Europe. Motulsky and his family entered Belgium shortly before the Germans invaded it. He was separated from his parents and siblings and spent the next year in camps across France, barely avoiding being sent to Auschwitz. He finally managed to escape Europe through Spain and Portugal to reach his father living in Chicago, where the rest of the family joined them several years later. Motulsky served in World War II and the Korean War, eventually earning his medical degree and learning genetics at Yale. In 1953, he began working at the University of Washington and, in 1957, set up the Division of Medical Genetics there, one of the first in the country.

That same year Motulsky published a paper in the prestigious *Journal of the American Medical Association.* In it, he pointed to a growing number of cases revealing peculiar "drug idiosyncrasies." Motulsky's prime example was derived from research conducted at Illinois's Stateville Penitentiary a decade earlier. Beginning in the 1940s, Alf Alving, a nephrologist with the University of Chicago, infected prisoners with malaria and then administered different synthetic antimalarial drugs to them to see what was most effective, part of a larger war effort aimed at protecting U.S. soldiers fighting overseas. Such research would never be approved today, but it was hailed at the time. In 1945, *Life* magazine published an essay about Alving's research with photographs of prisoners being bitten by mosquitoes and then suffering in bed with fevers and chills; the opening sentence praised the experiments where "men who have been imprisoned as enemies of society are now helping science fight another enemy of society."

The most promising antimalarial to come out of Alving's studies was primaquine. It was to be taken daily over the course of fourteen days and had relatively few side effects for the vast majority of people who took it. There was an exception, though. About one in ten Black men who took the drug developed hemolytic anemia, a dangerous condition during which red blood cells break down faster than the body can create new ones, leading to fever, weakness, dizziness, increased heart rate, confusion, and chest pain. Starting in the 1950s, Alving wanted to figure out why primaquine was unsafe for certain people and whether there was a safe dose that everyone could tolerate. In the first experiment, he and his team gave thirty milligrams of primaquine to 110 Black inmates; 5 of them developed anemia so severe that the scientists had to end the study prematurely for fear of killing the men. That was just the beginning of the ordeal for those 5 prisoners. Once they recovered, which took about two to three weeks, they were then put on a fourteen-day regimen of fifteen milligrams of primaquine to evaluate the impact of the halved dosage; 4 developed anemia, but it was milder. Once those same 5 men recovered yet again, they were started on a fourteen-day regimen of thirty milligrams of pamaquine, another antimalarial under development. All 5 became so anemic that this study too had to be prematurely halted. Based on the series of experiments, Alving and his colleagues concluded that fifteen milligrams was a safe dose for everyone (even those individuals with "primaquine sensitivity"), and the military subsequently began treating soldiers in the Korean War accordingly. In a follow-up study, Alving set out to determine why the small percentage of Black prisoners developed hemolytic anemia. The primaquine-sensitive prisoners, he found, had lower levels of glucose-6-phosphate dehydrogenase (G6PD), an enzyme that facilitated the proper functioning of red blood cells. Primaquine sensitivity thus became understood at the molecular level as G6PD deficiency.

Motulsky knew that other reports were emerging of some patients experiencing an unusual reaction to succinylcholine chloride. This drug was commonly used as an effective general anesthetic that lasted for just a couple minutes, but a small number of patients remained unconscious for hours, potentially requiring mechanical ventilation to keep them alive. Like primaquine sensitivity, differences at the genetic level were also discovered that could account for patients with this pseudocholinesterase deficiency. Motulsky, reflecting on these cases, called for greater collaboration between geneticists and pharmacologists, both of whom would benefit from the interaction: the geneticists would gain information about how people with different biochemical profiles responded differently to different drug exposures (the drug idiosyncrasies), and the pharmacologists would learn about the underlying molecular mechanisms that gave rise to these seemingly rare but potentially dangerous adverse reactions to their drugs.

Prisoner participants in the Stateville Penitentiary malaria studies.
Myron Davis/The LIFE Picture Collection/Shutterstock

Motulsky's 1957 paper is now commonly identified as marking the birth of pharmacogenetics—a term he did not use—the marriage of genetics and pharmacology, and he in turn is often called the father of pharmacogenetics. Friedrich Vogel, a prominent human geneticist from Germany, coined the term two years later, but it was really the Toronto-based pharmacologist Werner Kalow's 1962 book, *Pharmacogenetics,* that popularized the term and its associated idea. Kalow's treatise and his own research on the genetics of pseudocholinesterase deficiency didn't receive nearly the same amount of attention as Carson's *Silent Spring,* published that same year, but reporters and others were genuinely taken with the science. When Kalow presented elements of his book to the New York Academy of Medicine the same year it was published, *The New York Times* published both an editorial and an article about the event; a "reasonably simple chemical test," one journalist reported, could prevent the deaths of patients who might have an adverse reaction to commonly administered drugs.

Drug companies were less sanguine. Motulsky recalled telling a pharmaceutical executive he had found "a new way to find out about drug reactions. And he kissed me off: 'Drug reactions?'" There simply weren't that many cases of drug idiosyncrasies documented yet. Kalow counted only a handful in his book, and only a few more emerged into the 1970s. There were also financial forces at play. Starting in the mid-twentieth century, pharmaceutical companies invested heavily in the development of drugs based on a one-pill-fits-all business model. A company could either identify a very common ailment and develop a drug that everyone who had that ailment would take—the more common the ailment, the greater the sales potential—or wait for someone else to develop such a drug and then tweak the formula slightly to make it easier to take or produce fewer side effects. Many drugs that now line the shelves of your local pharmacy were developed during this period. Smith, Kline & French developed Tagamet for peptic ulcers, which was quickly followed by Glaxo's Zantac; for allergies, Parke-Davis developed Benadryl, on the heels of which came Hoechst Marion Roussel's Allegra and Schering-Plough's Claritin; Sanofi's blood thinner Plavix followed up on WARF's Warfarin.

Into the 1990s, however, dismissing Motulsky's advice became increasingly untenable for the pharmaceutical industry. Cytochrome P450 encompasses a whole family of enzymes (more than a thousand) that participate in the body's natural detoxification. These enzymes, found mostly in liver cells, target foreign substances in the body and tag them with extra oxygen atoms, which makes it easier to flush the substances from the body. The enzymes evolved to protect organisms from hostile chemical invaders, whether that be a toxin ingested in a poisonous berry by one of your ancient ancestors or a tablet of Allegra taken for a runny nose and watery eyes. They are both foreign substances to be tagged and removed. This makes understanding cytochrome P450 dynamics crucial if we want to add foreign substances to our bodies in the form of pharmaceuticals.

Many factors can impact how the detoxification process plays out inside us. Dietary factors include grapefruit juice, which inhibits P450 activity, thereby slowing down detoxification, and cruciferous vegetables like broccoli, which do the opposite. There are also naturally occurring differences between us that are mediated by our underlying biological makeup. In 1953, the same year that Alving's team was tracking down the molecular basis of primaquine sensitivity, James Watson and Francis Crick discovered the double-helical structure of deoxyribonucleic acid (DNA). The collection of all the nuclear DNA in an organism, its genome, is made up of millions or, in the case of humans, billions of nucleic acid base pairs of adenine (A), thymine (T), cytosine (C), and guanine (G). A gene is a segment of that DNA—a stretch of As, Ts, Cs, and Gs—which a cell transcribes into ribonucleic acid and then translates into a molecular product that contributes to the building and maintenance of the organism; a detoxifying enzyme is just one example of what that product can be. By the 1990s it was clear that there were dozens of cytochrome P450 genes that coded for an enormous number of P450 enzymes, and subtle differences in the nucleic acid sequences could have big effects on drug metabolism; someone with an A at one location where somebody else had a C—what geneticists call a "single nucleotide polymorphism," or a SNP (pronounced "snip")—could mean completely different cytochrome P450 profiles and, thus, completely different detoxification rates. These genetic differences in drug metabolism are spread out across different regions of the human genome (as opposed to being confined to a single gene), and so the terminology shifted to "pharmacogenomics."

Drugs for depression, high blood pressure, seizures, and blood clots, to name just a few ailments, were all found to be impacted by this naturally occurring genetic variation, and many of these were the blockbuster drugs upon which the pharmaceutical companies had built their businesses. Data was also emerging that indicated adverse drug reactions to blockbuster drugs were taking a huge toll. One study estimated that in 1994 alone more than 2 million hospitalized patients had serious adverse drug reactions and more than 100,000 had fatal ones.

It really wasn't until the 1990s, when "pharmacogenomics" became a common term, that the science behind it became essential to a health-care industry more and more invested in treating patients' ailments with pharmaceutical compounds. Incorporating pharmacogenomics into drug development and drug prescription seemed enormously promising. Pharmaceutical companies could design clinical trials that took genetic differences into consideration, which would then determine who got which drug; patients would be less likely to experience a dangerous or even deadly adverse drug reaction; and hospitals and insurance companies wouldn't have to absorb the costs of all the trial and error involved in identifying which drugs worked for which patients.

The pharmaceutical industry adapted accordingly. Some of the largest drug companies in the world formed a rare alliance. Pfizer and AstraZeneca, among others, along with a handful of academic institutions, pooled $45 million and formed The SNP Consortium, whose goal was to collectively identify 300,000 SNPs in the human genome that were associated with diseases and, thus, potential targets for pharmaceutical intervention. It was an enormous collaboration in pharmacogenomics.

Geneticists also began forming smaller start-ups to offer boutique genomic services to the drug giants. Gualberto Ruaño, in trim double-breasted suits and with a distinguished accent, was among the savviest. A child prodigy, he left Puerto Rico and spent the late 1980s and early 1990s split between the rarefied halls of Yale, in New Haven, Connecticut, where he earned his PhD, and the Connecticut State Police's Forensic Science Laboratory in Meriden, where he helped solve crimes. The two environments, seemingly worlds apart, shared an interest in incorporating insights from genetics. In the academic laboratories at New Haven, Ruaño studied molecular genetics, examining small segments of DNA in humans and chimpanzees that could shed light on the species' genomic similarities and differences; in the forensic offices at Meriden, he examined segments of DNA obtained from samples gathered at crime scenes and alleged perpetrators that could establish whether the samples came from the same source. After adding a medical degree from Yale to his credentials, Ruaño could see firsthand in his patients the dangers of adverse drug reactions. He soon decided to form a company that could bring the benefits of genomics into medicine. A lover of art history, Ruaño combined "genomics" with "Renaissance" and founded Genaissance Pharmaceuticals in 1997 to provide pharmacogenomic services to drug companies. Its Mednostics Program, for example, would help companies design clinical studies that took the genetic differences among research participants into consideration. Its Lazarus Program would work with companies to resurrect drugs that had failed regulatory hurdles in the past due to adverse drug reactions, using the power of genomics to distinguish those who would not suffer adverse reactions.

Ruaño needed wealthy investors who could supply Genaissance with the millions of dollars necessary to get the company up and running. That meant he required a business plan and a phrase that distilled the plan down to just a couple of words—words that someone with no background at all in genomics could understand, appreciate, and remember. "Pioneering Pharmacogenomics" sounded too academic. "It's a slogan," he recalled years later. "It's about messaging." So he opted instead for "Pioneering Personalized Medicine," and in so doing Ruaño introduced to the public a new way of characterizing the role of DNA in health care—genomics *personalized* medicine.

Ruaño and other pharmaceutical executives spread the word about the power of pharmacogenomics throughout the late 1990s, speaking with investors, reporters, and other scientists. They introduced to a wide audience a set of three marketing phrases that the pharmaceutical executives had been developing among themselves to sell pharmacogenomics: (1) pharmacogenomics was about delivering "the right drug, to the right patient, at the right time"; (2) pharmacogenomics worked in contrast to an outdated one-pill-fits-all pharmacology; and, most important, (3) pharmacogenomics "personalized medicine." Just as reporters showed great interest in Kalow's case for pharmacogenetics in 1962, science journalists and editors flocked to the buzz surrounding pharmacogenomics in the late 1990s. Between 1997 and 1999, a series of articles with titles like "Personal Pills" and "Just for You" appeared in newspapers and magazines, drawing largely on interviews with industry executives like Ruaño to explain to the public what was so exciting about the fusion of genomics and pharmacology. The stage was set, at the turn of the century, for a revolution in health care. All that was needed was the conclusion of the Human Genome Project.

. . .

It was past lunchtime, and the audience of teachers, students, and science enthusiasts were getting tired. They crowded into the chairs and along the aisles of the five-hundred-seat auditorium at the Smithsonian Museum of Natural History to celebrate the April 2003 completion of the Human Genome Project. Watson and Crick opened the ceremony to loud applause. But the subsequent three hours brought a lineup of federal scientists, academics, and policy advocates who overwhelmed the mostly public audience with talks full of technical jargon and dense biochemistry. Attendees slouched low in their seats, chins in hands, while those about the periphery leaned against the walls with arms folded.

Francis Collins was tasked with wrapping up. "You all must be feeling a little shell-shocked at this point," he commiserated with a grin. Any other scientist would have thanked the crowd for coming and bade them farewell, but Collins pulled out his guitar. Years ago, Collins explained as he casually slid bluegrass-style finger picks onto the digits of his right hand, anonymous volunteers donated their DNA to the Human Genome Project. Now that the human genome was sequenced and available to anyone who wanted to view it online, Collins asked the crowd to imagine one of those volunteers sitting down at home in front of their computer screen, staring at the string of

Francis Collins (*left*) with James Watson (*right*) at Smithsonian "Celebration of the Genome" kickoff event, April 13, 2003, a private event that took place two days before the public celebration.
The National Human Genome Research Institute

genetic letters, and wondering, "Is that me?" He started strumming his guitar to the tune of Del Shannon's 1961 hit "Runaway," and then he sang,

As I walk along the bases, in all three billion places, upon my PC screen,
Am I built for strong endurance, or loss of health insurance? Am I a mere machine?

I'm a'walking through the genes, don't know what all this means. No one else knows that it's me, behind that G and T.
And I wonder, w-w-w-w-wonder. Why, why-why-why-why-why I've got an A, you've got a C there. What does that say? Amazing DNA. D-D-D-D-DNA.

So I'm glad to know I've got some, of what Crick and Watson, found and brought them fame.
Use your own imagination, despite variation, we're really much the same.

By the final chorus, the whole audience was singing "D-D-D-D-DNA" and clapping in time. When Collins sang his last note, the crowd jumped to their feet for a standing ovation.

Collins grew up in the 1950s on what he called a "dirt farm" in the Shenandoah Valley of Virginia. Homeschooled by his mother until he entered high school two years early, he was encouraged to follow his own interests in his own fashion, which suited the inquisitive child. When he wasn't milking a cow or building a set for his family's local theater, Collins masterfully strummed his guitar with local bluegrass musicians. And when he wasn't doing that, he explored the natural world around him. He found the biological side of that environment, however, too chaotic and messy. The "overwhelming complexity of life," he sensed, lacked the sort of logic and predictability that his "budding reductionistic mind" favored, and so Collins gravitated instead toward the mathematical order of physics and chemistry.

The aspiring scientist went to Yale (after Motulsky but before Ruaño) to earn a PhD in physical chemistry. While there, though, he took a course in biochemistry in which he first encountered DNA. The genetic code, where nucleic acids bonded by the fundamental principles of chemistry and where a single mutation could generate a debilitating disease, astounded the graduate student. "Biology has mathematical elegance after all," he recalled. "Life makes sense." That reimagination of what the life sciences presented, it turned out, overlapped with him raising a young daughter, and the experience of fatherhood was profound. It shook him out of his solitary comfort crunching calculations that plotted the behavior of subatomic particles and inclined him toward a career that would allow him to interact regularly with people and have an impact on their lives. PhD in hand, rather than pursuing a career in chemistry, he decided to enter medical school and study the double-helical molecule that sparked his interest in biology.

Collins's pivot from protons to genes coincided with another major shift in his life. He was raised without religion and embraced a world without a deity. That gave way when he started meeting patients who faced death with a sense of peace and grace that both puzzled and inspired him. He began seeking cosmic order, where ethical behavior rested on moral laws just as reliably as the actions of nucleic acids rested on chemical laws, and where a divine mandate demanded altruistic consideration of others. Collins found in Christianity an evangelical worldview that was oriented around saving humankind—one that was in harmony with his growing professional ambitions. Insights from medical genetics, he thought, could allow him also to "contribute something to humanity" by combating the diseases that took a toll on his patients.

Collins, after completing his medical residency, joined the faculty at the University of Michigan. The lanky geneticist, standing well over six feet tall, with a no-frills combover of auburn hair and a matching chevron mustache, made a series of genetic discoveries there in the late 1980s and early 1990s that propelled him to fame. At the time, the Human Genome Project was just being conceived, so there was no reference map that geneticists could examine to hunt for genes. Instead, they had to slowly and laboriously assemble and then read through chunks of DNA, looking for tiny differences that separated patients with a disease from the general population. Teams from across the globe raced to home in on which chromosome, then which segment of that chromosome, and finally which precise point in the sequence of DNA a disease-causing gene resided. A geneticist joined the company of scientific elite with just one discovery. Collins made three in five years, pinpointing the genes responsible for cystic fibrosis, neurofibromatosis, and Huntington's disease, all debilitating and even deadly rare diseases. Such discoveries, for the born-again Christian, were like "moments of worship."

Around the time that Collins and his team were isolating the Huntington's disease gene, a shake-up unfolded at the NIH, which hosted the Human Genome Project. Watson, its director, ran into conflicts with the NIH administration and abruptly resigned just as the project was taking shape. Collins was a natural choice to take the reins. His stellar scientific credentials, abundantly clear by 1992, were part of the equation, but the NIH, in coordination with the Department of Energy, had launched the Human Genome Project as a massive, federally administered scientific endeavor, and the leader of that effort had to be someone who could solicit ongoing financial support from Congress, draw favorable media coverage, and excite taxpaying citizens. Collins fit the bill. He was fiercely competitive but rounded off the edges with a folksy charisma. He happily played his guitar in front of large audiences. He infused his scientific talk with religious inspiration. He rode a Honda Nighthawk 750 motorcycle to the lab and frequently posed on it for photographers.

But was the Human Genome Project the right move for Collins? He was, after all, doing very well at the University of Michigan, only in his early forties, and clearly hitting his scientific stride. Taking the position at NIH wasn't just about transplanting his family from Ann Arbor, Michigan, to Bethesda, Maryland. It was about reimagining his profession and his place on the scientific landscape—from the leader of a prestigious academic lab that could seek out the genes he found most medically important, to the administrator of a massive federal endeavor that would break up the human genome into smaller and smaller overlapping segments of DNA and then paste the whole thing back together using the areas of overlap as guides. Collins was hesitant, but he prayed and eventually saw it as a once-in-a-lifetime opportunity. "Here," he thought, "was a chance to read the

Francis Collins posing on his motorcycle. Ted Thai via Getty Images

language of God." He joined the NIH in 1993, taking over the Human Genome Project.

Collins proved himself an excellent choice as leader. As part of his hiring negotiations, he demanded that the NIH administration transition the National Center for Human Genomic Research to an institute, a more prestigious and more organizationally secure designation at the federal agency. In 1997, the center became the National Human Genome Research Institute. The timing was crucial because, soon after, the entire Human Genome Project faced an existential threat.

In the spring of 1998, Collins received a phone call from Craig Venter, a former NIH employee and even onetime collaborator on the Human Genome Project, who left his position with the federal government several years earlier to start a private company. Venter remained a familiar face within the genomics community, albeit one recognized as something of a renegade and a brash one at that. Collins was about to fly from Newark to Los Angeles, but Venter said he wanted to meet in Washington, D.C., so that they could chat in person. Venter was prickly and unpredictable by nature, which made the call unsettling and reason enough for Collins to begrudgingly accommodate the request by changing his West Coast flight to add a layover at Dulles. In a private room of United Airlines' Red Carpet Club lounge more commonly reserved for political lobbyists and business executives, Venter wasted no time on pleasantries. The Human Genome Project, he explained to the director of it, was in trouble; it was going to take too long, and it was mired by all the voices and red tape of the federal government. Venter and his company could do it faster, and they could do it better. Still, there was no reason their federal and private efforts couldn't work together, Venter assured Collins. He and his company would sequence the human genome. Collins and his team, Venter offered, could focus their sequencing efforts elsewhere. "You can do mouse." Collins sat dumbfounded. The nerve it took to spin such an insult as a collaborative opportunity was truly astounding.

News of Venter's company, Celera Genomics, and its plan to sequence the human genome before Collins landed on the front page of *The New York Times* on May 10, 1998. The threat to the Human Genome Project was obvious. Why waste taxpayer dollars paying for something that private industry was now going to do anyway? When Collins refused to play second fiddle by sequencing the rodent, a race ensued to see who could complete the task first. Reporters ate up the competition, dramatically characterizing Venter as a maverick genius on par with Galileo. The story quickly became one about the cumbersome, federal effort led by Collins struggling to keep up with the sleek, privately funded outfit run by Venter.

Collins navigated the pressure that followed with impressive political and media savvy. He decried Venter's "antagonism and excessive competition" as counterproductive to good science, but he also thrived in that cutthroat environment. To start, Collins publicly reframed the narrative of the race for the media. The Human Genome Project scientists were proceeding with the highest standards for data accuracy, while Celera was producing the "*Mad Magazine* version" of the human genome. The federal scientists were also committed to altruistically making the genetic information they produced free and available worldwide so that all humankind would benefit. Venter's effort, in contrast, selfishly lent itself to monetized access to the data. Then Collins cleverly bumped up the Human Genome Project's timeline for sequencing the DNA, initially scheduled to conclude in

2005, promising a publicly funded "rough draft" at the same time that Venter's company delivered its sequence, thereby neutralizing Venter's plans to cross the finish line first.

On June 26, 2000, Collins rounded a corner in the White House and began walking briskly toward the East Room. Suddenly somebody tugged on his sleeve. "I think I'm supposed to be out in front." It was the president of the United States of America. Collins fell in line behind President Clinton and alongside Venter as the three men walked in on a standing ovation from a packed audience of reporters, photographers, politicians, and scientists. The crowd gathered to officially declare the race between the public and the private efforts to sequence the human genome a tie. After the president called the product of the competition the "most important, most wondrous map ever produced by humankind," Collins took the lectern for his moment in the international spotlight, the first of many. The boy from Virginia still had the nasally twang of a rural bluegrass singer, but his hair was starting to show signs of gray. Collins smiled wide and held his chin high when he declared, "Today we celebrate the revelation of the first draft of the human book of life." There was extra reason to celebrate that day for Collins, and Venter didn't grin nearly so much throughout the ceremony, for the tie was, in reality, a win for Collins because it disincentivized Venter from continued commitment to genomic sequencing, having lost the ability to claim sole ownership of the genetic data. Venter, shortly thereafter, left Celera as the company shifted to investments in drug development. Collins and his Human Genome Project allies sequenced on. In April 2003, they announced that they'd finished sequencing the human genome $300 million under budget and two years early. The gathering at the Smithsonian that same month was timed to mark the historic achievement, to celebrate completing the human genome, and doing so alone, with Venter and Celera nowhere in sight.

Collins was also an effective leader of the project because, as a physician drawn to the science based on its clinical potential, he was passionate about its medical value and eagerly communicated that. "Despite variation, we're really much the same," he sang to the Smithsonian audience, but it was the genetic variation which made everyone different that most interested him. He sincerely believed that increased genetic information would revolutionize health care, and he told the audience that patients' personal genomes would be the key to solving health problems. Rather than treating all patients the same way, doctors could use genetic information to determine which cancer patients, for instance, would be administered chemotherapy, which would have only surgery, and which would be treated in some other manner.

But the promise of personalized medicine didn't stop with cancer, Collins declared from the stage. Genetics would revolutionize how physicians addressed all diseases, and the impact would go beyond life-extending treatments to pharmaceutical cures. In the next five to ten years, he expected the major genes for diabetes, Parkinson's, asthma, and many of the other most common diseases would all be discovered. These genetic findings would in turn completely transform the way those diseases were diagnosed and treated. By 2010, Collins foresaw a whole world of personalized medicine in which physicians tailored treatment plans and lifestyle changes to a patient's unique DNA. And by 2020, he said, "we will have a gene-based designer drug available for almost any disease that you can name." Back in 2003, it was hard not to get excited by Collins's vision, one where the completion of the Human Genome Project was a beginning, not an end.

But to find genes involved in common diseases, one human genome wouldn't suffice. Many were required so that differences in the genomes could be linked up with differences in health outcomes. By 2003, a number of other countries were putting together very large national research cohorts. Britain was launching UK Biobank, an attempt to recruit 500,000 Brits into a long-term study, and similar efforts were under way in Estonia, Japan, and Iceland. Collins thought the United States deserved its own nationwide study, and he was the natural figure to oversee it. That grand vision was typical of the geneticist's leadership style—a bold idea and a commitment to decisive action motivated by a calling to deliver on the untapped potential of genomic medicine. Finding such a project was particularly important for Collins because the Genome Institute's very existence became a matter of some discussion as April 2003 approached. It was created with the express purpose of sequencing the human genome, and that was finished. Wouldn't the reasonable thing be to shut down the whole institute now? Reframing the Human Genome Project as just the first step along a path toward making personalized medicine a reality, on the other hand, meant the Genome Institute would be a central cog at the NIH in perpetuity.

A few weeks after the Smithsonian celebration, Collins conveyed that to his staff at the Genome Institute. "We haven't had much discussion about the longitudinal, large-scale cohort study lately, but I continue to think that this is a project that needs to be pursued (in the U.S., not just in the UK, Estonia, Iceland, and Japan)." The challenge was finding some way to pay for such an endeavor. He'd just spent billions of taxpayer dollars on the Human Genome Project, so immediately asking Congress for billions more was risky, even for a scientific celebrity in the middle of his victory lap. Funding from pharmaceutical companies was one possibility, he proposed. "Is there a way," Collins also wondered, "to connect this with the National Children's Study?"

The National Children's Study, as ordered by Congress in the Children's Health Act of 2000, was run out of the National Institute of Child Health and Human Development (NICHD), housed in the very same NIH building as the Genome Institute. That made Duane Alexander, director of Child Health, the leader of the whole enterprise. Alexander grew up in Annapolis, on the Maryland coast of the Chesapeake Bay, and spent virtually his entire life in the state. The only exception was when he headed north to attend Pennsylvania State University in the late 1950s and brought his father's tuba with him, playing in the Penn State marching band the very same instrument that his father played in it decades earlier. Alexander intended, at the height of the Cold War, to study aeronautical engineering and "send rockets to the moon," but, like Collins, he became drawn to the gratification of working with people, so he switched to a medical track.

After graduating, Alexander returned to Maryland, completing his medical school training, residency, and a fellowship in pediatrics all at Johns Hopkins University. There, he had the opportunity to study under Robert Cooke, who was both head of pediatrics and a pioneer in reimagining the way that the medical world treated children with developmental disabilities. Cooke had two children of his own with developmental disabilities. In an era when the norm would have been to institutionalize them, Cooke raised his daughters at home with his other children and focused on surrounding them with a nurturing social and medical environment. Through a relationship with Sargent Shriver and his wife, Eunice Kennedy Shriver, bonded by their shared commitment to improving the lives of children with developmental disabilities, Cooke also became both pediatrician to the Kennedy family and one of the most prominent voices on child health policy in the Kennedy administration. Among his lasting contributions was persuading the president to create a new institute at the NIH devoted to children's health and development.

Cooke's stature and mentorship proved life altering for Alexander. While at Johns Hopkins, Alexander specialized his pediatric training in developmental disabilities, and it colored the way he approached medical research and medical care. In contrast to Collins's aversion toward biological complexity and his preference for reductionistic order, Alexander embraced the causally tangled world that led to developmental disabilities and that had to be carefully considered when treating children with the diagnoses. The nature of the conditions required it because the causes of developmental disabilities were so diverse and complicated—a chromosomal aberration as an embryo, a viral infection during fetal development, a complication

Duane Alexander (*left*) with Senator Ted Kennedy (*right*) during the senator's visit to the National Institutes of Health on April 3, 1996. *NIH Record*

during delivery, a toxic exposure as a newborn, an impoverished social environment as a toddler, some combination of these. President Kennedy signed the legislation creating the Child Health Institute in 1962, just as Alexander was coming back to Maryland to enroll in medical school. After his fellowship and barely into his thirties, Alexander joined the relatively new institute as a scientific adviser in 1971. Fifteen years later, in 1986, Alexander was named director of the institute his mentor first envisioned.

Alexander's ascension and longevity at Child Health were due in no small part to the fact that he was adored by everyone around him. The man was cherublike—short, bald, on the heavy side, always cheerful. He was the Platonic form of a pediatrician, the kind of person who would only smile a bit wider as a baby peed on his tie. Alexander dressed up as Santa Claus for holiday events at the institute, and the suit fit. He also inherited that passion for mentorship from Cooke, reflected as a resolute loyalty toward and open embrace of his whole staff. In contrast to Collins's migration from rural isolation to suburban life, the pediatrician went in the opposite direction, leaving Annapolis behind to raise his family on a farm in Ellicott City, more than thirty miles northeast of the NIH campus. Alexander grew fruits and vegetables there and brought the harvest in for staff. When two of them were married, he gifted mason jars of his homemade ketchup and tomato sauce. Special occasions at the institute were marked with shipments of ice cream from his alma mater Penn State's famous Berkey Creamery.

Alexander infused the National Institute of Child Health and Human Development with his unflinching optimism. "Duane could fall off a cliff," one longtime collaborator surmised, "and halfway down say, 'Well, so far so good.'" He cultivated and patiently waited for consensus to form around big decisions, sometimes putting off controversial choices that seemed urgent to others, knowing that such scenarios often resolved themselves of their own accord. Some found this management style frustrating and wishy-washy, but it clearly served him and Child Health well. By 2000, when Congress passed the Children's Health Act, the sixty-year-old was already among the longest-serving institute directors in NIH history.

The Child Health director brought that optimism and inclusive mindset to his leadership of the National Children's Study. The mandate for the study, as laid out in the act, with its call for a complete assessment of so many influences on children's health, essentially demanded it. Experts from a wide variety of backgrounds had to be enlisted to serve on two dozen working groups that helped guide the shape of the study. The creation of these groups, it turned out, brought an additional bureaucratic layer. According to the Federal Advisory Committee Act, if working groups provide input, then a federal advisory committee open to the public must also transparently process all their input. As a result, the National Children's Study Federal Advisory Committee was formed. At the same time, the Children's Health Act also called for the whole study to be guided by an Interagency Coordinating Committee with representatives from the NIH, the EPA, and the Centers for Disease Control and Prevention (CDC). By 2003, between the working groups, the Federal Advisory Committee, and the Interagency Coordinating Committee, there were several hundred federal and academic scientists from across the country working for or advising the National Children's Study, all funneling ideas and recommendations up through the various channels to Alexander.

The range of personnel created as many challenges as it did safeguards. The plan for the study was for researchers to go into participating children's homes, collect environmental data, and administer tests. Opinions about what data should be collected and which tests should be administered, though, were diverse. The list of planned tests and data grew longer and longer, which conflicted with practical considerations about what it would be reasonable to expect of the families that participated and what could be funded. Alexander, true to form, tolerated the tension in exchange for buy-in from a range of disciplinary stakeholders.

Despite approving the study, Congress initially allocated no federal dollars to pay for the National Children's Study, leaving it in an unfunded mandate status. But because the experts in disciplines ranging from developmental psychology and family studies to pediatric endocrinology and environmental toxicology believed in its value, they were willing to advocate for its importance at their home institutions and at their professional scientific meetings, which maintained an enthusiasm until funds emerged to launch. In the spring and summer of 2003, Child Health began preparing a presentation with this goal in mind, intending to request financial support from President George W. Bush.

That's when Collins reached out. On July 23, Collins and his deputy director, Alan Guttmacher, walked the short distance from their Genome Institute over to Child Health, where Alexander and the program director of the National Children's Study, Peter Scheidt, awaited. There was a formal meeting space in Alexander's office with a conference table and stiff seats, but Alexander invited Collins and Guttmacher to join him and Scheidt more comfortably in a sitting area with a long couch and cushioned chairs. Collins, on that pleasant summer afternoon, came to Alexander with a request: to piggyback off the National Children's Study by recruiting the parents and grandparents of the kids enrolled in Alexander's project and including them in Collins's large study of adults. The proposal had something to offer both directors and both institutes. Collins's side could focus on the genes and adults, while Alexander's community attended to the kids and the environment. Collins got to latch onto a federal project that was already approved by Congress, while Alexander got to bring on board one of the most prominent scientists in America. The whole conversation lasted no more than twenty minutes. The scientists shook hands and agreed to move forward together.

After the Genome leadership walked out the door, Alexander and Scheidt debriefed. Scheidt, for his part, had never met Collins before; he found the famous geneticist warm and charming. Alexander, on the other hand, had spent a decade in gatherings of institute directors with Collins dominating conversations and angling for initiatives that benefited the Genome Institute. It was indeed a unique opportunity to leverage Collins's celebrity into the funding that they needed. But Collins was also notoriously pushy when it came to getting what he wanted. Still, the trade-offs were worth it for the optimistic pediatrician, although they'd have to proceed with caution. "Every time Francis gets an idea," Alexander said with a chuckle, "it ends up costing me."

Collins, back at the Genome Institute, contacted Elias Zerhouni, director of the NIH, to whom both Collins and Alexander reported.

Collins wanted the agency to get behind his piggyback idea, and that was ultimately the NIH director's decision to make.

The NIH is the largest biomedical research institution in the world. Luke and Helen Wilson, heirs to department store and manufacturing giants, donated their sprawling Tree Tops estate in Bethesda to the federal government in the 1930s. The property sat on Rockville Pike, a major artery running from the Maryland suburbs into Washington, D.C., and the Roosevelt administration saw the land as a perfect location for medical research that required both state-of-the-art laboratory space and agricultural terrain large enough to accommodate the vast number of animals included in the studies. Just six structures were initially perched on a rise in the landscape, but the agency grew in the ensuing decades to include more than seventy buildings spread out across three hundred acres. There are now more than two dozen institutes and centers affiliated with the NIH, staffed by more than twenty thousand scientists, physicians, and administrators. There is, in addition to the Genome Institute and Child Health, an institute devoted solely to cancer care, and another focused on eyes; one for mental health and one for addiction, to name a few. The grounds today feel like a bustling university campus, minus the swarming population of undergraduates; Georgian-style architecture of red brick and white columns sit along meandering paths that loop around rolling hills. Only the tight perimeter security set up after 9/11 gives away the fact that it's an arm of the U.S. government and not Bethesda's local state college.

Each institute of the NIH operates independently, with its own director, its own staff, its own initiatives, and its own budget derived from taxpayer dollars. Overseeing the whole agency is the NIH Office of the Director, located in Building 1. The director is nominated by the president and often changes with each new administration. Because of this structure, the NIH director usually has the most direct access to the White House. That access can be the difference between getting presidential support for an expensive new initiative or not, and that presidential support can be the difference between an expensive new initiative launching or dying a premature death.

Zerhouni, nominated by President Bush in 2002, was attracted to the collaborative proposal from Collins and Alexander. After he arrived at the NIH, one of his first priorities was finding initiatives that fell through the organizational cracks of the federal agency, either because of the way that biomedical research tended to get siloed within institutes or because of the way a particular project might seem too big for a single institute. The American Family Study, as Collins and Alexander called it, presented a nice opportunity to fill one of those cracks if the costs over time were affordable in the context of tightening federal budgets. Zerhouni, Collins, and Alexander presented the material to White House staff in September 2003, and then, in November, the president's domestic policy adviser, Margaret Spellings, invited them to come back a second time because there was genuine interest within the administration. The ideal outcome would be for President Bush to announce support for their study in his next State of the Union address, scheduled for January 2004.

While Collins and Alexander waited for news from the White House, they organized a scientific gathering in December designed to assess the merits and challenges associated with the piggyback proposal at the heart of the American Family Study. Fifty scientists descended on the Hyatt Regency in Bethesda over the first three days of December 2003. The generic hotel conference room was an unremarkable environment in which to brainstorm a remarkably ambitious biomedical research endeavor for the United States. The participants, sitting around a rectangle of tables littered with water bottles and PalmPilots, were a mix of National Children's Study affiliates, personnel from the Genome Institute, and other scientists who had experience with designing and running large-scale and long-term human studies. As with all scientific workshops, the professional dress code ranged from a crisp suit and tie to a wrinkled polo shirt and jeans. Collins stoked excitement, asking, "Can we afford NOT to do something like this?" The task for the group was to judge whether it was feasible.

The participants identified a number of challenges that the American Family Study designers were going to face if they built off the National Children's Study. If a necessary condition for participation was having a child in the National Children's Study, that meant individuals without children (and health conditions specific to them like infertility) would be systematically excluded. Logistically, keeping track of three generations of a family would also be difficult; all generations do not always live in the same area. On the other hand, if entire families were recruited, there would be a family-level commitment that incentivized the continued participation of each family member. Collins also figured the 100,000 children would bring with them roughly 160,000 parents and 240,000 grandparents (assuming not all parents and grandparents were available). That added up to 500,000 participants, which would match Britain's nationwide effort.

When Collins led the last session devoted to "Next Steps," all signals indicated the collaboration was going to proceed rapidly. Geneticists from the Genome Institute would meet with the pediatric scientists on the Interagency Coordinating Committee. A list of hypotheses that the family study could test would be worked up. Focus groups would be convened to test out marketing the nationwide effort. But in early January the NIH scientists learned that President Bush was not going to mention the American Family Study in his State of the Union address. Instead, he would announce his support for NASA sending humans to Mars.

Collins and Alexander, without presidential support, had a decision to make. They could still appeal to Congress; after all, it had called for the National Children's Study, and the American Family Study was an extension of it. They could also seek out an audience with the secretary of health and human services, who oversaw all health-related agencies in the federal government. Cabinet secretaries can prioritize initiatives of their own.

The December meeting devoted to assessing the American Family Study did identify logistical and scientific challenges associated with the piggyback design; however, none of those issues were insurmountable. When it came to missing issues of infertility, Collins noted that simply recruiting an additional sample of adults without children could alleviate that problem. And if the study was truly national in scope, then researchers in Portland, Oregon, could interact with children and parents living there while researchers in Miami collected data from the grandparents there. But the brief collaboration revealed deeper sources of tension that could not be so easily resolved. Even though the two sides agreed on the revolutionary potential of a nationwide study of health in America, they could not find common ground concerning the nature of the revolution that they were seeking.

Collins wanted to use the information to usher in a new era of gene-based clinical care. For him, the American Family Study was essentially a massive genetics study, the next big initiative for his Genome Institute. When he visited the White House in the fall with Alexander and Zerhouni, Collins framed the whole exercise as the natural follow-up to the Human Genome Project. And when the geneticists crafted text that President Bush could read at his State of the Union address, the initiative was characterized as a father-son affair: Bush 41 launched the Human Genome Project, and now Bush 43 was bringing the health impacts of personalized medicine to all Americans.

The National Children's Study started from a very different place. It was an extension of Carson's *Silent Spring,* the environmental justice movement, and the pediatric community's call for attention to the unique environmental threats posed to kids. The study was meant to address America's health-care crises by reorienting medicine in America around a public health model that identified toxic threats in children's environments and removed them.

Fear of getting eclipsed by Collins also materialized. His request to piggyback off Alexander's study of kids started to feel like an offer to piggyback off *Collins's* adult study. The proposal was a massive add-on to the National Children's Study that would fundamentally alter its scope and trajectory. And yet almost all planning for it had been quickly swept up by the Genome Institute. All the weekly meetings were held at Collins's institute; the invitations for the December meeting were issued from Collins's institute; the charge to attendees at that meeting came from Collins. Alexander was not anti-genetics. Far from it. During his time as director of Child Health, in fact, the pediatrician was a major advocate of implementing prenatal and newborn screening. But those technologies were largely about identifying rare genetic conditions, not preventing the development of common illnesses, which was the focus of the National Children's Study and necessarily demanded prioritizing attention to children's surroundings. A cautionary proverb circulates within the NIH community: "There are never mergers for Francis. Only acquisitions." There was genuine concern that Collins's proposed merger was becoming an acquisition, and the environmental focus of the project would get lost in his repackaging of the enterprise as Human Genome Project, Part 2.

Guttmacher and Scheidt spoke by phone on the morning of January 14, after which Guttmacher updated Collins: "Their approach is to go back to Congress to build necessary support for [the National Children's Study]." Guttmacher told Scheidt that the Genome Institute planned to request an independent review of their idea for a nationwide study focused on adults, and that would delay any effort on their part in seeking financial or political support for several years, "which would allow [the National Children's Study] to fight for Congressional support without our jumping the queue."

Scheidt conveyed a similar message back to Child Health: "I just talked to Alan Guttmacher. Genome is, for the time being, holding on their plans for an adult cohort." Omission from the State of the Union address, it seemed, created an opportunity for Collins and Alexander to amicably part ways.

2

The End of a Partnership

IN ONE OF THE ICONIC IMAGES of World War II, the leaders of the United States, the U.K., and the Soviet Union sit side by side in Crimea in February 1945 at the Yalta Conference, where they met to discuss the administration of postwar Germany and Europe. On the right, Premier Joseph Stalin sits rigid and stern. On the left, Prime Minister Winston Churchill, holding his ever-present cigar, grins

Winston Churchill, Franklin D. Roosevelt, and Joseph Stalin at the Yalta Conference in February 1945. Franklin D. Roosevelt Presidential Library & Museum

through squinted eyes. In the middle, President Franklin D. Roosevelt slouches, looking frail and haggard. It wasn't just the war weighing on him. He was dying. Two months later, while resting in Warm Springs, Georgia, with family and friends, the president complained of a sudden headache and slumped over. He was dead within hours, having suffered a major cerebral hemorrhage. The news came as a shock to the American people, but it wasn't entirely a surprise to the physicians monitoring him. For some time it was clear that Roosevelt was in the late stages of cardiovascular disease.

President Roosevelt was no isolated case. America in the mid-twentieth century was in the midst of a cardiovascular disease epidemic. Cardiovascular disease results from the buildup of plaque on the walls of arteries, making them narrower, harder, and less efficient at circulating oxygen-carrying blood. Eventually blood clots form and wreak havoc; if a clot blocks blood flow to the heart, a heart attack results; when a clot blocks flow to the brain, it's a stroke. Even without blood clots, the weakening circulatory system can exhaust the cardiac muscle, leading to heart failure. In the 1940s, cardiovascular disease was the leading killer of Americans, responsible for one-third of all deaths. FDR's death was just the galvanizing case.

In response, Roosevelt's successor, Harry S. Truman, signed the National Heart Act in 1948. It added the National Heart Institute to the NIH and tasked it with running a study that addressed the deadly epidemic. (It also resulted in a name change; the "National Institute of Health" became the "National Institutes of Health.") Physicians in the 1940s had few options for treating cardiovascular disease once it manifested itself, especially at the late stages, so the scientists designed an innovative program focused on prevention rather than treatment. If they could identify the factors that increased the risk of developing the disease, then physicians could help their patients before they became sick. With that goal in mind, the federal researchers descended on the town of Framingham, Massachusetts, just west of Boston, in 1948 and recruited roughly fifty-two hundred of the town's twenty-eight thousand residents. The adult participants had to show no signs of cardiovascular disease upon enrollment and agree to check back regularly for the life of the study, which was anticipated to be twenty years. In fact, Congress has continued funding it to this day. The scientists recruited the children of the original participants as well as the children's spouses in 1971, and the grandchildren were added in 2002, allowing the scientists to track how conditions passed down from generation to generation.

The Framingham study forever changed the way physicians approach heart health. Much of what doctors now know about the risk factors associated with cardiovascular disease comes from Framingham. High blood pressure, the study confirmed, is a risk factor for cardiovascular disease, as are smoking, obesity, and physical inactivity. Indeed, the very term "risk factor"—a concept that pervades medicine today—is widely credited to the study's scientists.

When in 2003 Francis Collins and Duane Alexander were collaborating on the American Family Study, Framingham was very much on their minds. Collins even took to calling it "Framingham on steroids!" The heart study, though, despite its tremendous impact, had important limitations. The original participants were almost entirely white and middle class, and the study was, as its name made clear, a heart study. It was not designed to investigate cancers, developmental disorders, birth defects, respiratory diseases, visual impairments, or any of the other medical conditions that afflicted people. "Framingham on steroids!" meant a much larger sample of participants, nationwide representation, more diverse participants, and more health conditions investigated.

. . .

The Collaborative Perinatal Project had been formed in the 1950s in response to a terrifying trend impacting America's youngest. As the postwar baby boom peaked, pediatricians began noticing an increase in developmental disabilities such as cerebral palsy, vision and speech impairments, reading and learning deficiencies, and cognitive delays. Twenty million Americans were thought to live with one of those debilitating conditions. There was also a shockingly high rate of infant mortality in the nation, and most frightening, it was unclear why children were dying or developing disabilities at those rates. Were they due to complications during labor and delivery? Exposure to something during fetal development? Environmental encounters in infancy? With no clear understanding of the causes of the health problems, there was no reliable way to combat them.

The federal government responded with the Collaborative Perinatal Project, which recruited pregnant mothers across the United States and tracked their health, and the health of their babies, for several years into childhood. Like the Framingham study, the project was entrusted to the NIH. In this case, it was the National Institute of Neurological Diseases and Blindness that administered the program, but the recruitment, retention, and data collection were handled by twelve academic institutions spread out across the nation. Between 1959 and 1965, recruiters enrolled roughly sixty thousand women into the most ambitious birth cohort study in America to date.

Janet Hardy was in charge of the effort at Johns Hopkins University and the diverse Baltimore community it served. She was born in British Columbia, and her father was a doctor. After he told her, "No daughter of mine is ever going to be a physician," she became one, revealing a tenacity early in life that never left. Hardy reached the Johns Hopkins Hospital's pediatric program in the early 1940s and helped set up the first neonatology ward there. She then spent most of the 1950s working for the Baltimore City Health Department's Bureau of Child Hygiene, focusing on preventive medicine for the city's youth. But in 1957, when Hardy had the opportunity to lead one of the twelve centers embarking on the Collaborative Perinatal Project, she returned to Johns Hopkins. Navigating the academic medical world as a woman in the 1940s and 1950s was daunting; she was only the fifth woman to reach the rank of full professor at the medical school at Johns Hopkins. Hardy was sharp, diligent, and extremely organized. "Put It in Writing," said the large sign over her desk, a reminder both to document everything and to turn research ideas into published products.

When studying environmental impacts on children's development, Hardy and the community of pediatric epidemiologists knew full well that the key was collecting high-quality, objectively measured data over, and over, and over again. Children, from conception through fetal development through infancy and then into childhood and adolescence, develop at a remarkable pace. That is all happening in response to an ever-changing environment—the womb and mother, the home and family, the neighborhood and friends, the school and peers. To understand that process and the factors impacting it for better and worse, scientists need to repeatedly gather carefully documented data about the children and their environments so that patterns can be identified and linked up with the health outcomes.

Hardy and her team met with the Baltimore-area children frequently: multiple times while their mothers were pregnant; at four, eight, and twelve months during infancy; and then at ages three, four, seven, and eight. The researchers collected maternal serum during and right after pregnancy so that they could analyze what was reaching the fetuses. They performed a battery of neurological examinations to track the children's sight, hearing, and language development. They collected information about the families' socioeconomic situations, and they interviewed the parents about their own health and the home life in which the children were raised.

Hardy was obsessive about staying in touch with the four thousand or so mothers and their children spread out across Baltimore who were participating in the project. On Christmas and at birthdays, cards were sent to their homes that contained return postage; that way, if a family moved, the card would be returned, and Hardy's team could track down the new address. Staff were sent to the neighborhoods to speak with family, friends, grocery store clerks, to find out where the family went. Hardy, on other occasions, would send taxis to the homes of the mothers and children in order to bring them in for the regularly scheduled visits. When the mothers were leaving, Hardy always told them, "Thank you for coming. Without your help and interest there would be no study."

Accumulating all the necessary data was costly—both in terms of dollars (about $100 million, which is close to $1 billion in today's value) and in terms of time investment—but it proved well worth it. The project revealed that cerebral palsy was caused by events during pregnancy, not during labor and delivery. It showed that it was perfectly healthy for women to gain weight while pregnant—indeed, healthy for mom and healthy for baby. It exposed dangers for children born to adolescents, sparking widespread efforts to prevent teen pregnancy. It offered guidance for how to prevent brain damage in babies with jaundice. It linked exposure to rubella in the womb with subsequent birth defects. It also set standards for hospitals across the country offering neonatal care to their newborns. For every one thousand babies born in the United States in the 1950s, nearly thirty never reached their first birthday; by the 1980s it was down to fewer than thirteen.

The Collaborative Perinatal Project ended in 1976. The data was available to researchers afterward and continued to provide valuable information about the healthy development of children in the United States. But no new efforts were made in the United States to create another massive, longitudinal study of the nation's newborns since those Collaborative Perinatal Project babies were enrolled. As the twentieth century was ending, skyrocketing rates of asthma, obesity, and diabetes—chronic conditions that risked a lifetime of health problems for the nation's youngest generation—scared America's pediatricians. So it was in 1998 that a group of specialists in developmental disabilities were sitting around in Alexander's office at the NIH brainstorming ideas for the next pediatric initiative when one of them said, "Well, it's been forty years since anyone has tried to conduct a large pregnancy cohort study in the United States." Alexander looked over to one of his top epidemiologists. What did he think of the idea? "We've joked about this," he replied, "but nobody has ever had the raw nerve to ask for the billions it would take to do it." Maybe it was time to try. The group nodded in agreement and, just like that, the National Children's Study began to take shape.

. . .

Humans have been interested in understanding their own heredity for millennia and putting that information to use: making predictions about the traits of children based on the traits of parents, trying to select the traits of babies who are born, providing guidance in light of the patterns. The Talmud, for example, warns that if two baby brothers die of excessive bleeding after circumcision, then subsequent brothers as well as maternal cousins should be exempt from the tradition, documenting both the dangers of what would later be identified as hemophilia and the unique, maternal pattern of its inheritance. In sacred Hindu texts, we are told that what siblings have in common is due to their father, while the points of difference are attributable to their mother; readers can also find advice on how to increase the chances of having a son. The Hippocratic Corpus from ancient Greece notes the hereditary nature of illnesses like "a spleen disease" and temperaments like being "phlegmatic."

The study of human heredity took both a scientific leap forward and a moral leap backward at the beginning of the twentieth century. In 1900, research conducted by the Austrian monk Gregor Mendel decades earlier on thousands of pea plants was recognized to shed important light on patterns of inheritance. The discipline of genetics was born. Researchers across the globe began breeding fruit flies and maize to see how traits passed from generation to generation. The experiments indicated that something physical was transmitted—what geneticists soon called a "gene"—and it was located on chromosomes in the nuclei of cells; the investigations also allowed for tracking how traits like eye color in flies or seed texture in peas arose in each new generation by virtue of which genes were inherited on which chromosomes.

Some champions of this new science attempted to extend the practical lessons of animal and plant genetics to human biology and in ominous ways. If the purple eyes of flies and the texture of peas followed straightforward hereditary patterns, they thought, then surely human traits like intelligence and criminality did too. These eugenicists feared that people they deemed "unfit"—those who lived in poverty, who were convicted of crimes, who lived with cognitive disabilities—were outbreeding people they deemed "fit," those who scored high on intelligence tests, who held positions of prestige, who were wealthy. Eugenics was all about reversing that perceived reproductive threat. Since the eugenicists understood human heredity to follow the same rules as animal and plant biology, the challenge of increasing the prevalence of creativity in each new generation or decreasing the prevalence of poverty was about multiplying the transmission of creativity genes or restricting the transmission of poverty genes. In the darkest chapter of American genetics' history, eugenicists obtained legal and legislative victories in the early decades of the twentieth century: curtailed entry of foreigners from entire regions deemed biologically unfit; involuntary sterilization laws granting states the ability to take away the reproductive rights of more than sixty thousand Americans deemed unfit to have children.

Fortunately, eugenics gradually lost its scientific credibility throughout the middle of the twentieth century. It became clear that eugenicists were packaging racism, ableism, classism, and nativism in the language of biology. There were no simple genes for poverty or criminality, so no amount of sterilization or immigration restriction would eliminate the traits that eugenicists found undesirable. The real eye-opener was when it became clear that the Nazis' effort to exterminate Jews and people with disabilities was inspired by American eugenics, and so the horrific culmination of the eugenic vision became plain for all to see when concentration camps were discovered across Europe.

The stigmatization of eugenics, however, didn't mean that people suddenly stopped caring about human heredity. The branding and focus just shifted. A new crop of "heredity clinics," departments of "medical genetics," and programs in "genetic counseling" appeared with the same focus on understanding and controlling human heredity. Arno Motulsky set up one of the early divisions of medical genetics at the University of Washington in 1957, the same year Olga Owens Huckins found her dead songbirds in Duxbury and Janet Hardy took charge of the Collaborative Perinatal Project in Baltimore. The big difference between the new genetic organizations and the older eugenic varieties pertained to how the biological information would be utilized. Eugenicists wanted to use the information to guide who did and didn't procreate. Medical geneticists, on the other hand, aspired to empower prospective parents with information that guided their own reproductive decision making.

That shift created a practical challenge for medical geneticists. To make genetic information useful for individuals, the science had to say specific things about specific people. It wasn't enough to say "cancer runs in your family" or "since your father died early from a neurodegenerative disease, you are at increased risk of dying early from the same thing." Individuals knew that from their lived experience. They needed scientists to tell them whether one family member was likely to develop colon cancer or whether another one was safe from Huntington's disease. That clinical motivation, alongside the discovery of

DNA as the physical basis of heredity, led geneticists to seek out the specific genes—the exact stretches of DNA in the human genome—that were responsible for medical diseases and disorders.

Geneticists' primary tool for gene hunting in the mid- to late twentieth century was *linkage analysis,* a reference to the fact that genes which are physically close to one another on a chromosome tend to be inherited together more often than genes that are far apart. Armed with maps of known genes' locations, geneticists set out to find families that were in the unfortunate position of harboring rare diseases that turned up in each new generation. By comparing genetic markers of those who inherited the disease and those who did not, medical geneticists were able to home in on the genomic region that differentiated the sick from the healthy. Throughout the 1970s and 1980s, geneticists worked alongside large families in Venezuela, the Netherlands, the United States, and other places to identify the genes for diseases that haunted those families, such as Huntington's disease, hereditary retinoblastoma, cystic fibrosis, and breast cancer. That information allowed geneticists to tell specific family members whether they carried a dangerous variant.

Geneticists refer to Huntington's disease and cystic fibrosis as "rare Mendelian traits" because they are among the unusual human features that tend to play by the same hereditary rules as the peas that Mendel studied, where there was a fairly straightforward gene-trait relationship. If you have the genetic variant, you get the condition; no genetic variant, no condition. Those Mendelian traits were indeed rare. Geneticists, however, hoped the linkage analysis method could be extended to conditions that impacted far more humans, like cardiovascular disease, depression, and asthma—what geneticists call "common complex traits." These traits don't follow the same hereditary patterns as Mendel's peas, which makes it more difficult to locate and track the genes associated with them. Geneticists learned that in the shift from rare Mendelian traits to common complex traits, individual genes make a far smaller contribution. As a result, linkage analyses simply didn't have the biological resolution necessary to find those subtler genes hiding behind tiny effects.

Neil Risch and Kathleen Merikangas, two genetic epidemiologists at Yale University in the early 1990s, were well aware of the problem facing their discipline. Even after Risch left Yale for Stanford, the two of them continued corresponding about the state of affairs in human genetics. They wrote "The Future of Genetic Studies of Complex Human Diseases," which appeared in *Science* in 1996, suggesting that human geneticists shift away from linkage analyses and embark on enormous association studies. In those studies, a scientist gathers up a large group of people with some disease (the "cases") as well as a large group of people without that disease (the "controls") and then seeks to determine what all the cases have that the controls lack. Risch and Merikangas proposed seeking out the source of the disease difference between the cases and the controls in the minute differences between the genomes of the two groups. If a catalog of points of variation in the human genome could be accumulated, then geneticists could look to see if any of the differences at the DNA level correlated with the differences in having the disease. To make the whole thing work, Risch and Merikangas figured geneticists would need to get information on hundreds of thousands of different places in the human genome where the single nucleotide polymorphisms resided and document the different forms they took. They realized in 1996 that the Human Genome Project was setting the stage for precisely that; once that project strung together the full sequence of one human genome, it could be used as the foundation for seeking out the places in it where there were differences among people's DNA.

The genetics community came to call Risch and Merikangas's proposed approach to studying common complex traits "genome-wide association studies" (or GWAS). The two scientists advised trying it with up to 1,000 cases and controls, but a sample of 10,000 humans would find genetic signals that 1,000 did not, and 500,000 would find effects that 100,000 did not.

Genome-wide association studies at the turn of the twenty-first century set genetics on a quite different methodological trajectory from the environmental science that Hardy applied with the Collaborative Perinatal Project. In the child health study, the challenge was carefully collecting and tracking environmental variables over, and over, and over again, and determining which environments to measure, how to measure them, and when to measure them over the course of a child's development. Hardy came to appreciate the logistical problem as well—how to recruit and keep engaged children and their families who would allow scientists to collect intimate information about them very often and for so long. For genome-wide association studies, the challenge was technological. Some differences in the DNA would correlate with some differences in the complex common traits, so that all the geneticists needed was a bit of blood from the participants and medical records would show who developed a given disease and who didn't. This was made all the easier as medical records migrated to electronic formats around that same time, allowing for health data to be searchable and quantitatively analyzable. The thing preventing genome-wide association studies from including 100,000 or even 500,000 participants was the costs and time of genotyping them.

. . .

On the face of it, Alexander and Collins both wanted to improve health in the United States and agreed that a very large, diverse, long-lasting, nationwide study of the American people was a time-tested approach to addressing the nation's health crises. That's why the collaborative American Family Study seemed like such a natural union. But, as we saw, within months it became clear that they were not on the same page about which health revolution they were facilitating. Was the study about taking the lessons of the Human Genome Project and transforming health care into personalized medicine, or was it about taking the lessons of the environmental health movement and making the places around us safer? How one answered that question involved both a vision for the future of health in the United States and a scientific road map for how to reach that destination. The destinations were not the same, so the paths would be divergent.

The Collaborative Perinatal Project set the standard for what a large, longitudinal birth cohort study could achieve in the United States, and it was very much on the minds of the child health scientists once Congress called for a new nationwide study of kids in the Children's Health Act. Peter Scheidt, program director of the National Children's Study, knew the Collaborative Perinatal Project well. In fact, he worked alongside Hardy for a period of time in the 1970s, utilizing data from her project for research of his own.

Alexander and Scheidt didn't know it at the time, but they first crossed paths in 1956 when Annapolis High School faced off against Howard High for the Thanksgiving Day football game. While Alexander was playing tuba for the Annapolis band, Scheidt was on the field carrying the football as starting halfback for Howard. Both born in Maryland, Scheidt moved around a bit more but always found his way back to their home state. After completing his undergraduate studies at Johns Hopkins University, he earned his medical degree at Duke, completed an internship at the University of Utah, and worked for two years with the Indian Health Service in Oregon and Arizona. Then it was back to Maryland for his residency at Johns Hopkins just as Alexander was departing the same program, followed by three years with the Indian Health Service again, this time on the Alaska Panhandle. Scheidt was back in Maryland working for the U.S. Food and Drug Administration in the 1970s, and that's when he and Alexander first met by way of their shared interest in the safety and efficacy of phototherapy for treating newborns with jaundice. It's also when Scheidt had the opportunity to interact with Hardy, utilizing some of her Collaborative Perinatal Project data to compare how kids in the 1970s who received the treatment compared with children from her study who didn't get it years earlier. When Congress passed the Children's Health Act in 2000, Alexander needed somebody to join the Child Health Institute and serve as director. Scheidt, he learned, was available, and when Alexander offered him the position, Scheidt happily accepted.

The high school football player retained his athletic form and competitive nature well into adulthood. Compact and fit, Scheidt moved with a purpose. On Wednesday afternoons, he rushed out of the program office and drove east for forty-five minutes until he reached Annapolis, where he, his wife, and their children sailed in boat races on the Chesapeake Bay. Temperamentally, though, he was Alexander's twin, and they enjoyed each other's company. Scheidt's home, on the property where he grew up, was right on Alexander's route into Bethesda from his farm, and the Child Health director occasionally picked up his National Children's Study director on the drive to NIH. They were both products of Robert Cooke's pediatrics program at Johns Hopkins, quick to smile, and inclined toward an egalitarian approach to management. That made the two pediatricians of one mind when it came to leading the National Children's Study, erring on the side of a "let a thousand flowers bloom" approach to the input they received from the hundreds of scientists who gave advice by way of the study's working groups, Federal Advisory Committee, and Interagency Coordinating Committee.

The plan for what data was going to be collected, how it was going to be gathered, and how often it was going to be accumulated was ambitious. Specially trained staff were going to roll up to the children's homes in large vans, which contained an assortment of different devices created to grab samples of air, soil, water, and dust. Interviews designed by family and child psychologists were going to be administered with the parents and, when the children were old enough, them too. Videos would record parent-child interactions and the caretaking environment. Data was also to be collected from the children's schools and their neighborhoods—things like automobile traffic and the prevalence of graffiti. Participating families were to be visited at least fifteen times: three times during the mother's pregnancy; at birth; at one, six, twelve, and eighteen months after birth; at the child's ages of three, five, seven, nine, twelve, sixteen, and twenty. The Collaborative Perinatal Project made it clear, however, that the scientific payoff would be well worth it. If the twenty-first-century National Children's Study could do for asthma and autism what its twentieth-century predecessor did for cerebral palsy and jaundice, it would transform the health landscape of a nation.

The source of the most internal strife in the National Children's Study was the question of how the children were going to be selected for recruitment (or "sampled"). Researchers can go probability based or convenience based. A probability sample is designed to give every member of the target population (for example, children born in the United States) a chance at being recruited into a study; the population is divided into units (for example, counties, neighborhoods) with each unit guaranteed a chance at representation in the study population; units are randomly sampled, and then potential participants are randomly selected from those units. A convenience sample, as the name suggests, draws the sample from the target population at places that are most convenient to the task, such as a hospital.

A convenience sample is easier to generate, but what you can legitimately infer from what you learn about the people in the study is limited. If you recruit only at hospitals, the results of the research cannot apply to the entire U.S. population; it will only generalize to those who had access to such clinical care. For the social and behavioral scientists who were helping with the National Children's Study, the sorts of complex phenomena that they knew impacted children's health—racism, economic opportunity, educational resources, family dynamics—required a probability sample in order to be able to reliably say that the results generalized to the entire nation. "You have to drop our topic," one social scientist explained, "to drop our way of getting the sample." The question was whether the probability sample was logistically and financially feasible.

A series of contentious professional consults, panel reports, and straw polls took place between 2002 and 2004, with sides becoming increasingly entrenched. Critics of the probability sample warned it wouldn't work, while advocates of it cautioned that the convenience sample would mean losing the support of social and behavioral scientists. Ultimately, it was Alexander's decision to make. The inability to reach a consensus, after nearly two years of trying, pushed the mild-mannered pediatrician outside his managerial comfort zone, but he was still guided by the principle of being as inclusive as possible. The Children's Health Act called for a *complete* assessment of the physical, chemical, biological, and psychosocial environmental influences on children's well-being, and so Alexander reasoned that the probability sample had to at least be tried. If it proved too unwieldy, the National Children's Study could always fall back on a convenience sample. The plan was to team up with academic medical centers across the nation—following the lead of the Collaborative Perinatal Project—and task them with taking charge of the recruitment, retention, and data collection.

The National Children's Study was conceived as a response to the child health crises facing the nation at the turn of the twenty-first century. In addition to autism and asthma, the focus was on addressing obesity, schizophrenia, congenital heart defects, attention deficit hyperactivity disorder, diabetes, traumatic brain injuries, and other diseases and disorders. Those conditions were having a huge impact on the health of the nation's youth. Specific hypotheses regarding the diseases of interest were identified in order to probe specific scientific questions: What role does glucose metabolism in pregnant women play in birth defects? Do prenatal infections increase the risk of neuro-developmental disabilities? What dietary exposures associate with asthma? How do family resources impact a child's developmental health trajectory?

The study designers believed a sample of roughly 100,000 children could capture enough of those cases for study to allow for testing the hypotheses under consideration. They also intended to follow the children until they were twenty-one so that the exposures and experiences of childhood could be traced to the health outcomes of adulthood.

The nationwide study sought by Collins at the Genome Institute was inspired by the desire to remake health care in America in the mold of personalized medicine—to tailor patients' clinical care to their DNA. That placed Collins's deputy director, Alan Guttmacher, squarely in the middle of the effort. Guttmacher, in certain respects, was the polar opposite of Collins. In contrast to Collins's dirt-farm upbringing, Guttmacher was raised in a family of prestigious physicians. His father was a forensic psychiatrist who testified in Jack Ruby's trial. His father's twin brother, an obstetrician-gynecologist and Alan's namesake, was a leading advocate of reproductive health issues and the president of Planned Parenthood. His mother, also a psychiatrist, was a prominent human rights advocate and the first female dean of students at the Massachusetts Institute of Technology.

Guttmacher's family might have naturally driven him to seek out the sort of fame that Collins achieved, following his siblings into psychiatry and health-care administration. Guttmacher, however, only reluctantly became a physician. While he was still in high school, he lost his father to leukemia and inferred from that experience that a career in medicine amounted to watching people slowly die. The Harvard grad tried his hand at farmwork, middle school teaching, and assisting political campaigns. It was only later, at his uncle's memorial, that one of his brothers told Alan to stop fighting it and join the family business—medicine. Still, when Guttmacher finally went back to Harvard to attend medical school, his goal was to be a family physician in a rural town out of the public eye.

In another respect, though, Collins and Guttmacher were quite similar. They were both passionate about genetic research and its potential to change patient lives. When Guttmacher finished his medical training as a pediatrician, he moved to Vermont, where he led the Vermont Regional Genetics Center. He also obtained an academic appointment at the University of Vermont, where he could conduct pediatric research of his own. In the late 1980s, Guttmacher's assistant informed him that "some geneticist" was going to be passing through town and hoped to meet. Guttmacher had to look up information on the guy to make sure he was legitimate. The name he searched under was "Francis Collins." Collins was closing in on his string of remarkable discoveries and not famous yet, but Guttmacher still happily met with the up-and-coming geneticist and learned that they shared an interest in hereditary hemorrhagic telangiectasia, a rare Mendelian trait that results in vascular malformations that can lead to dangerous bleeding. The two stayed in touch, even as Collins shot to fame and moved on to lead the Human Genome Project.

In 1999, as the Human Genome Project was heading toward completion, Collins began looking ahead to the medical implications of the work. He advertised for a clinical adviser to join the Genome Institute, to which Guttmacher applied and took the job. Once Guttmacher arrived, he quickly moved up the administrative ranks to become Collins's deputy director. Beyond expertise and degree (they were among the few figures with the Genome Institute at the time with a medical degree), there was the Collins-Guttmacher dynamic. Guttmacher, a pediatrician like Alexander and Scheidt, displayed a quiet and unassuming demeanor. Collins and Guttmacher made an effective pair—the relentless showman next to the gentle pediatrician. When Collins could be overwhelming, Guttmacher was by his side with a calm voice and a reassuring smile.

The duo proved particularly effective when it came to surrounding themselves with the expertise necessary to design a nationwide study of their own, apart from any collaboration with Child Health. They quickly formed working groups, arranged a series of planning meetings, and drafted up a proposal, essentially seeking to accomplish in just a couple of months what it took several years to accomplish for the National Children's Study. The result looked very little like the final product developed by the child health scientists. The geneticists planned only to collect blood and urine at an initial visit, conduct a physical exam to get measures like height and weight, and ask the participants to fill out a questionnaire regarding physical activity, dietary habits, and smoking/alcohol use. Researchers would follow up with the participants in four years, although some telephone and email communications were planned in between.

The sample sizes of the two studies were also radically different. Collins wanted to recruit 500,000 participants—five times what the National Children's Study hoped to include. Following the turn prompted by Risch and Merikangas toward larger and larger data sets, Collins felt they needed a population that large if they were to pick up the very subtle genetic effects that contributed to common complex traits. What's more, Collins could afford to invest in a much larger sample because far fewer resources were going to be devoted to things like home visits, costly environmental measurements, and frequent participant interactions—the hallmarks of the National Children's Study.

Collins, for his part, also steered clear of tying his study to explicit hypotheses. Science evolves, the thought went, and so too do scientific hypotheses. As a result, the study envisioned at the Genome Institute was conceived primarily as a data-gathering platform that collected information that could bear on current or even future hypotheses proposed by scientists. The timeline for the genetics study was about a decade—a year to get the thing up and running with initial recruitment, followed by a couple of checkups at four-year increments, and then a year to wind down and analyze the data. So it was going to last half as long as the National Children's Study.

These two studies could easily have coexisted in the United States. The purpose and design of each were unique, and so the deliverables for the American people were distinct. The National Children's Study aimed to improve the world so that kids were less likely to need a trip to the hospital; success for the genetics study would amount to better clinical care in the form of tailored treatments and prescriptions for adults. These were the natural divisions of labor when Collins and Alexander tried working together on the American Family Study, and the agreement was that they'd stay in their own lanes after they broke off the collaboration in January 2004.

Almost immediately after the two went separate ways, however,

Collins made a fateful choice that forever altered the relationship between the two efforts and, in turn, the trajectory of biomedical research in America. Not content to sit back and wait to see the fate of the National Children's Study, he decided to add kids to his study too, and also advertise it as providing a balanced investigation of both genetic *and* environmental contributions to health. Those decisions ensured that the project Collins was developing would become a direct competitor to the National Children's Study, meaning they would have to fight for the same financial and political resources.

Collins asked to chat with Alexander, and the two institute directors spoke over the phone on February 12, 2004. The National Children's Study, Collins told Alexander, was in "deep trouble"; it was taking too long to get up and running without congressional funding, and rumors were spreading that Alexander's inclusive attitude toward scientific input was making things cumbersome for the project. Collins could do it faster, and he could do it better. Still, there was no reason the two couldn't continue to collaborate, Collins assured Alexander. The Child Health director could study the kids born to the adults in Collins's project.

There was a definite irony to how Collins formulated his new proposal for Alexander. He took the insulting script that Venter ran on him five years earlier and flipped it onto the Child Health director. "You can do mouse" became an encouragement for Alexander to let go of the National Children's Study along with all the work that hundreds of scientists had put into it for several years, and pick up whatever children trickled into Collins's cohort over time. Collins's request just months earlier to piggyback off Alexander's study really had morphed into an offer for Alexander to piggyback off Collins's project. The merger had become an acquisition.

Alexander informed Scheidt and the other members of the Interagency Coordinating Committee from the CDC and EPA about Collins's proposal for a new model of their study of kids, and there was universal agreement that the idea was a bad one. There would be too few children enrolled in Collins's approach to test the hypotheses pertaining to child health and development, and they would join the project much more slowly. Equally important, the carefully and repeatedly measured environmental data on which the National

Children's Study was centered would be deprioritized compared with Collins's focus on the DNA of the participants.

That left Collins in a bit of a bind. He wasn't trying to, Venter-style, leave the NIH, form a company, and embark on a privately funded nationwide study that would compete with the publicly funded effort in Child Health. He was trying to create an alternate, publicly funded effort at the Genome Institute within the same federal agency that housed Child Health. The official version of events for the public and for the elected officials in Congress who controlled funding for the agency had to be that everybody at NIH was playing well together.

Collins, as a result, shifted to plan B. If Alexander couldn't be co-opted, then perhaps he could be "convinced not to oppose." On March 2, Collins met Alexander in person. Collins towered over Alexander—both physically and in terms of personality—so it was surely an intimidating interaction for the Child Health director. Collins once more warned Alexander about the fate of the National Children's Study, but Alexander continued to express his commitment to it. Collins, as a result, told Alexander that he was proceeding with the plan that Alexander had refused the previous month—a national study of his own involving both child and adult participants that purported to investigate genetic and environmental contributions to health equally. Collins described it as a "backup plan" to Alexander's National Children's Study. The problem with that characterization was that a backup plan required allowing the primary plan to proceed first. Collins later debriefed Guttmacher about his testy exchange with Alexander, frustrated by his fellow institute director's refusal to give up on the National Children's Study. "This is a bit messy," he admitted, "but I don't know how else to proceed—unless [the NIH director, Elias Zerhouni] is prepared to tell Duane to abandon the [National Children's Study], which I don't think is likely."

Collins next focused on packaging the genetics study so that it would seem obvious it was the best option for a nationwide study in the United States. The easiest way to do that was in the name: the American Gene Environment Study, or AGES. Both the full name and the acronym conveyed the purported advantages of Collins's model in comparison to Alexander's. The National Children's Study, the Genome Institute director explained, focused on the environment; AGES, Collins promised, was about treating both DNA and the environment equally, even though one glance at AGES revealed that it didn't pay nearly as much attention to environmental impacts on health as it did to genetic influences. Meanwhile, Collins said his study included children, adults, and older Americans, and would therefore be a benefit to more age-groups than a study of just youth.

When Collins opted for a competitive relationship in 2004, decades of environmental research was having a profound and beneficial impact on the health of Americans. Studies of the relationship between lead and neurodevelopmental problems were followed up with abatement programs that decreased the amounts of lead in children's blood. Research that linked up smoking with lung cancer was joined by programs aimed at preventing smoking and followed by significant drops in deaths from cancer.

The geneticists' personalized medicine, on the other hand, was already showing signs of serious limitation in 2004. The first sign was the recognized need to switch from linkage analyses to genome-wide association studies in the 1990s. Risch and Merikangas characterized the association studies optimistically as the "future" of genetic studies of complex traits, but implicit in that shift was the acknowledgment that human genetics' future was about chasing down smaller and smaller genetic effects. Smaller effects, in turn, meant less obvious routes to make genetic predictions or interventions on the genes in a way that improved health outcomes for the entire population.

That tacit concession was joined by a real shock that came out of the Human Genome Project. Going into that international collaboration, geneticists thought humans had about 100,000 genes—a gene or two associated with depression, a couple for obesity, another for dementia. But it turned out that humans have only about 20,000–25,000 genes. This means the human genome behaves in all sorts of complicated ways that human geneticists simply didn't anticipate and, again, makes it far more difficult to translate into interventions and predictions.

Nonetheless, drug companies were still drawn to pharmacogenomic interventions even if effects on improved health were small. Biotechnology companies knew that to make those pharmacogenomic interventions work, doctors and researchers needed fast and cheap access to people's DNA; that attracted enormous investments in genotyping and gene sequencing technologies. The media were fascinated by the promises of genetic "breakthroughs" and "miracle cures," which made reliably flashy headlines. And both conservatives and progressives in Washington, D.C., found elements of personalized medicine to be politically attractive. This now became a question of money, politics, and influence.

3

Industry Relationships

SMELTERTOWN FIRST TOOK SHAPE along the banks of the Rio Grande in the 1880s. For the Mexican American residents who made up most of the population—more than three thousand at its peak—it was "La Esmelda." There was a theater, a grocery store, a YMCA, a church, and elementary and vocational schools. It was a vibrant community of families who lived there for generations, oftentimes in the same homes that were passed down from grandparent to parent to child. The people were historically, culturally, economically, and spiritually tied to that stretch of land on the outskirts of El Paso, Texas. If you visit the location now, you'll just see a sign that reads "Smelter Cemetery, 1882–1970" and graves marked with wooden crosses flaking white paint or mounds of rock worn smooth by the relentless wind and sand, the final resting places of the residents who called Smeltertown home for nearly a century. The people and buildings are all gone because Smeltertown was razed to the ground in 1973.

To understand Smeltertown and its ultimate demise, you need to start with the enormous smelter around which the town grew up. Robert Safford Towne built a lead smelter along the Rio Grande in the 1880s, and the American Smelting and Refining Company (Asarco) purchased it in 1899, expanding the operation to smelt copper and zinc too. The Guggenheim family, famed for their subsequent philanthropic contributions to arts and aviation, controlled Asarco for

Smeltertown Cemetery circa 1970. © *El Paso Times*—USA TODAY NETWORK

the first half of the twentieth century, producing their tremendous wealth. For decades, Asarco was among the world's largest producers of metal, operating more than a dozen smelters across the United States.

Humans have been smelting for millennia, but the industrial process became big business in the nineteenth century. Once raw ore is mined from Earth and reaches a smelter, it is crushed and then heated in giant furnaces along with chemical reagents that help to separate out the desired elements from the unwanted materials. The liquefied metals that result are poured into molds and eventually sold. The by-products—sulfur dioxide, arsenic, mercury, lead, slag—are often released into the air or dumped onto the surrounding land.

As Asarco's El Paso smelter grew, so did demand for employees to work it. Many families across northern Mexico were familiar with Asarco because of its mining operations in their region. The promise of good-paying work just across the Rio Grande in Smeltertown lured thousands of immigrants north to El Paso in the early decades of the twentieth century. The company and the Mexican American workforce were mutually dependent, but there was an obvious power imbalance reflected in the racially segregated neighborhoods. The white managers, engineers, and scientists of Smelter Terrace lived in large, single-family houses made of brick and wood with yards along tree-lined streets. Their homes had electricity and plumbing, and the families had access to company amenities—tennis courts, a bowling alley, ice skating on frozen cooling ponds in the winter. On the Smelter Hill side, the Mexican American families occupied tenement apartments made of concrete block and adobe. Their small homes had no electricity or running water; one public lavatory was available for men, another for women. According to Monica Perales, whose family was from Smeltertown and who wrote a history of the people and place, the Mexican American families tolerated this racial divide in exchange for the steady income offered, along with the services that Asarco reliably provided its workforce: potable water from a pump house, a fire and police force, a hospital, and a company store.

The Mexican American residents of Smeltertown made the most of the racially segregated world in which they found themselves. They opened their own taverns and restaurants. They danced at the YMCA and worshipped at the Catholic church. When in 1933 Father Lourdes Costa stared at nearby Cerro de los Muleros (Mule Drivers Mountain) and envisioned a massive cross there, his congregants committed to making it a reality. They first hauled lumber up the steep incline to install a giant wooden cross. That was soon replaced by an iron version. The residents constructed a path up to the peak so that pilgrimages could be made safely. Finally, in 1940, a sandstone statue of Christ standing nearly fifty feet high was set in place. Cerro de los Muleros became Sierra de Cristo Rey. The land on which Smeltertown rested belonged to Asarco, but the history and community and sense of family and place belonged to the Mexican Americans who made La Esmelda their home.

Throughout the 1960s, external forces crept in on Smeltertown, both literally and figuratively. El Paso, throughout the early and mid-twentieth century, expanded rapidly. As a result, Smeltertown eventually found itself in a very different geographic relationship to the city. What started as an industrial complex on the outskirts of town eventually became a neighborhood swallowed up by the metropolis. Alongside the physical encroachment, there was the growing social awareness that emerged throughout the 1960s concerning the environmental and health dangers posed by industrial activities.

Asarco tried to keep those forces at bay. A towering smokestack from which the smelting by-products were spewed was built in 1951; in 1967, the company raised it to more than eight hundred feet, visible from miles around. The effort was an exercise in the industrial motto "dilution is the solution to pollution"; if you sent the toxic arsenic and lead higher into the atmosphere, the thought went, then it would travel farther and simultaneously become less concentrated. When those substances finally fell back down to Earth, they would be too diffuse to be unhealthy. For the families of El Paso that came to reside in neighborhoods right next to the smelter, however, that assurance was unconvincing. They saw the yellow clouds of sulfur dioxide belch from the smelter and blow their way, and they started complaining.

El Paso city health officials, responding to the public outcry, launched an investigation and determined that between 1969 and 1971 alone Asarco released more than 1,100 tons of lead, 560 tons of zinc, 12 tons of cadmium, and 1 ton of arsenic into the air. Soil studies confirmed that there were unsafe amounts of toxic substances in the surrounding area. The quantities alone led the city and the State of Texas to sue Asarco. Still, there were so many unanswered questions. Could the dangerous materials found in the ground be linked to the smelter? What was the geographic distribution? Did the chemicals have any impact on the health of people being exposed? The El Paso officials contacted the Center for Disease Control in 1971, and the CDC agreed to send one of its elite Epidemic Intelligence Service teams.

Investigating a lead smelter was a novel undertaking for CDC personnel in 1971. Until the previous year, the federal organization was known as the Communicable Disease Center, reflecting its historical focus and expertise concerning infectious diseases. The Epidemic Intelligence Service teams were trained as rapid response units that could descend on a measles or mumps outbreak, identify the source, and contain the spread. Lead was something else entirely. The head of the service teams, though, knew he had a young, Harvard-trained physician from Boston Children's Hospital who'd treated children with lead poisoning back in Massachusetts. That was good enough to put Philip Landrigan in charge of the Smeltertown investigation.

Landrigan was just a few years out of medical school when he was given twenty-four hours to pack up and fly to El Paso. For the tall and athletic pediatrician from New England who'd never heard of Smeltertown, the dry and dusty terrain of a Texas border town was otherworldly. Landrigan made things familiar, drawing on his CDC training and treating the challenge as an infectious disease outbreak. He and his team drew up concentric rings emanating out from the smelter in one-mile increments. By day, they collected blood samples from the adults and children, dust and soil from the neighborhoods; at night, they paid a penny to cross the bridge over to Mexico and have dinner in Ciudad Juárez. The results, when they took the samples back to CDC headquarters, were frighteningly clear. Lead levels in children's bodies and in homes in and near Smeltertown far exceeded amounts deemed safe by the federal government; these toxic concentrations then gradually decreased the farther away from the smelter one went.

Connecting the smelter to the toxic lead was the first step. The second was showing that the health impacts of lead exposure followed that same distribution. Landrigan returned to Smeltertown intending to conduct a series of studies on the children around the smelter, to see whether those children with the highest lead levels in their blood had associated cognitive impairments. When he returned, however, he learned quickly that Asarco was not sitting idle and letting the CDC control the situation. Company officials argued that any exposure in the area was due to preexisting lead in the soil and dust, not from their smelter. The company purchased an ad in an El Paso newspaper assuring the residents that there were no health concerns to worry about, but they would generously continue to monitor things for precautionary purposes. Asarco's scientists also approached the El Paso Board of Health and offered to conduct a health impacts study of their own, which they promised would be equivalent to the CDC's plan and would be paid for with private dollars; that way the taxpayers of Texas wouldn't have to foot the bill for costly science. The Board of Health accepted the offer and told Landrigan that his invitation was rescinded. Since the Epidemic Intelligence Service officers operated at the request of local and state officials, that meant the research was canceled.

Boxed out, Landrigan told his team to sit tight. He boarded a flight from El Paso to Austin, where he managed to arrange an audience with the Texas attorney general. John Hill heard the young scientist out and then promised him in a firm drawl, "Doctor, don't worry. The way will be made straight." When Landrigan returned to El Paso, his health study was back on. It confirmed the severity of the threat. Children with elevated lead levels had significantly lower IQs and significantly slower reflex reaction times than children who did not carry the same lead exposure burden.

Asarco ultimately settled with El Paso and Texas, agreeing to pay a series of fines, install additional emissions control equipment, and pay medical expenses for the children with elevated lead levels. But more important, Landrigan's research made national news and changed the conversation about particulate lead released into the air and the devastating health risks associated with exposure to it. Smelting companies could no longer dismiss concerns about toxic substances around their factories as attributable to other sources. Lawsuits soon followed in Tacoma, Washington, and Kellogg, Idaho, aimed at the smelters in those communities.

Sadly, this story did not have a happy ending. Asarco, by settling, got to keep operating its smelter. To avoid future problems, the residents of Smeltertown, which was on company property, would have to relocate, their neighborhoods leveled. The Mexican American families protested forcefully. They wanted a solution that protected their children but simultaneously allowed them to stay in the homes that their grandparents occupied, in the community with which they identified, and in the physical place that had come to represent their local history and culture. They even offered to buy the land and go through the process of rezoning it from industrial to residential. To no avail. The injustice at Smeltertown, in hindsight, was a textbook example of environmental racism, but that concept and attention to it were still a decade off. The residents received eviction notices in October 1972 and had to be gone by January 1973.

Razing Smeltertown, while solving Asarco's immediate legal problem, by no means eliminated the health threat in the region. The smelter continued to operate until 1999, only shuttering when world copper prices dropped. Elevated lead levels continued to be detected in both El Paso and Ciudad Juárez for decades after Landrigan conducted his research.

Landrigan's experience with Asarco was not unique. Carson's *Silent Spring* in many ways set the stage for the dynamic that emerged between environmental health researchers and the industries that they studied. Just as Carson's alarms about the dangers of DDT were dismissed by the chemicals industry, so were Landrigan's warnings about the harm done by lead written off by Asarco. Herbert Needleman, in the 1970s and 1980s, conducted studies that showed how even low levels of lead exposure posed health risks for children and thus required drastically reducing the federal limits on lead emissions and exposures. The lead industry, in response, tried to discredit Needleman; they persuaded his employer, the University of Pittsburgh, to open a research misconduct investigation probing whether he'd falsified his data. Naomi Oreskes and Erik Conway, in their gripping book, *Merchants of Doubt,* document how this industrial response to research that threatens the bottom line has been deployed against all sorts of environmental and public health problems for decades—tobacco products, acid rain, ozone depletion, climate change.

Carson, Landrigan, Needleman, and many others provide examples of just how ugly things can get when private industry deems certain science to be a threat. But it's crucial to understand how a science can be shaped by its relationship to industry even if things don't become combative. Scientific research in the United States has become increasingly dependent on private industry for funding; as a result, industry indifference can be as influential as industry antipathy. This goes for scientific research generally, and especially for health research. In the decades after World War II, the federal government made a concerted effort to become an international leader in science and funded more than half of all health research conducted in the United States. But starting in the 1970s, the government contracted that investment while private companies simultaneously expanded their financial influence. In the 1990s and the first decade of the twenty-first century, private industry—pharmaceutical and biotechnology companies in particular—funded close to 60 percent of all health research, while the federal government funded less than 40 percent, and nonprofit foundations contributed just a few percent. The trend has only accelerated in the years since, with industry now financing close to 70 percent of U.S. health research.

This transformation has been part of a concerted effort to commercialize science in America—to shift science from a place where it is first and foremost a public good to one where it is first and foremost a private commodity. Federal and state legislators throughout the 1970s, 1980s, and 1990s, inspired by neoliberal faith in the universal power of market competition, put laws in place that financially incentivized researchers to direct their science toward profit making. Intellectual property protections were enhanced so that scientists and their institutions that made discoveries stood to benefit financially from the insights. Research universities now typically house some form of technology commercialization office to help their faculty navigate the legal and bureaucratic process of translating their experiments from ideas into valuable products. Even the National Institutes of Health, that crown jewel of publicly funded biomedical research in the United States, couldn't avoid the trend. Congress, in 1990, established the Foundation for the NIH, an organization created to raise private dollars that then funded and shaped public research at the federal agency.

Landrigan, in the ensuing decades, rose from a young member of the Epidemic Intelligence Service to become the father of environmental pediatrics, an internationally acclaimed authority on environmental epidemiology, a routine provider of expert testimony before Congress, and the media's go-to voice on the dangers of lead exposure, as well as asbestos, pesticides, and unsafe drinking water. His trajectory reveals just how different the experience is for a scientist who lacks industry support compared with a scientist who is industry funded. The latter generates an idea or makes a discovery that someone recognizes as having the potential for profit, and a private company invests resources in bringing that idea to reality. An environmental health scientist like Landrigan, whose research has little financial appeal for industry, is typically supported by either a government entity (for example, the CDC, the EPA, a local public health department) or a nonprofit (for instance, the March of Dimes); the results of the research are then used to, for example, shape some new legislation designed to protect public health or serve some lawsuit designed to compensate people harmed by a company's practices. The 1993 report that announced the key insight that children are not just tiny adults was *Pesticides in the Diets of Infants and Children*, the product of a committee chaired by Landrigan. Its support came from the National Academy of Sciences, a nonprofit. Its impact, the Food Quality Protection Act of 1996, set new federal standards for pesticides that were most likely to reach the mouths of kids.

It should come as no surprise, then, that in 2000, when the Children's Health Act called for a nationwide study of children designed to investigate the environmental impacts on their health, Landrigan and Needleman together advocated for the creation of the National Children's Study. And when the first sites of recruitment were selected randomly and announced in 2005, Landrigan applied to lead one of the most logistically challenging locations.

. . .

In 1848, two cousins in their twenties left Germany and set sail for the United States. One was a chemist, the other a confectioner. With a $2,500 loan from family, they set up shop in Brooklyn. The chemist developed medicinal compounds that were in high demand, like antiparasitics; the confectioner figured out how to make the bitter treatments more palatable by adding sweet concoctions, like almond-toffee flavoring. In 1849, they incorporated under the name of the chemist: Charles Pfizer & Company.

Throughout the nineteenth century, Pfizer expanded to become a leader in the chemicals industry. It produced iodine and morphine for Union forces during the Civil War. In the 1880s, Pfizer refined the process of manufacturing citric acid. That development coincided with a nation suddenly crazed with soft drinks, like Coca-Cola and Dr Pepper, a main ingredient of which was citric acid. In the 1920s, Pfizer scientists started producing citric acid using molasses rather than lemons and limes imported from Europe, freeing up reliance on international suppliers. The company also developed a new fermentation process that allowed for mass-producing penicillin. When Allied forces invaded Nazi-controlled France in 1944, most of the penicillin that arrived on D-day was made by Pfizer.

John Smith, who helped lead Pfizer through the war years, made a pivotal decision for the company in 1950. He was dying of cancer when he suggested to his successor fundamentally altering Pfizer's business model. Pfizer, a chemicals company, turned a profit by manufacturing materials for other companies to use in their own products. But at that moment Pfizer had a new antibiotic ready to roll out, Terramycin. Smith was said to have advised from his deathbed, "Don't make the mistake we made with penicillin and hand it over to other companies. Let's sell it ourselves. Go into the pharmaceutical business if we have to." Terramycin was the first product that carried the Pfizer name on its label and was sold directly to pharmacies and hospitals. It was an enormous financial success, and it announced a new, major player in the pharmaceutical industry. Pfizer, throughout the 1950s and 1960s, expanded to become an international powerhouse, with branches throughout Central America, South America, Europe, and Asia.

Pfizer, like large pharmaceutical companies in the second half of the twentieth century, built its fortune on blockbuster drugs. Terramycin, a one-pill-fits-all medicine, was a broad-spectrum antibiotic that performed effectively against more than a hundred different infectious bacteria; indeed, farmers could use it on their pigs and cattle. The company invested heavily in research and development, rolling out a series of products throughout the 1980s and 1990s that became go-to drugs for the health-care industry. Feldene, a nonsteroidal anti-inflammatory used to treat arthritis, was the first to generate $1 billion in profits for Pfizer. Glucotrol was among the most prescribed treatments for diabetes. Pfizer's Zoloft—alongside Eli Lilly's Prozac—treated depression and anxiety. Other successes included the antifungal Diflucan, Procardia for high blood pressure, the antibiotic Zithromax, and Zyrtec for allergies. By the mid-1990s, Pfizer was counted among the largest pharmaceutical companies in the world and brought in billions of dollars in revenue each year.

Pfizer, despite all its success, was at the forefront of embracing pharmacogenomics. The strongest point in favor of incorporating pharmacogenomics into drug discovery and development pertained to avoiding adverse drug reactions in patients. This movement really gained steam in the 1980s and then the 1990s, when research on cytochrome P450 revealed all the ways that genetic differences impacted those enzymes crucial to drug metabolism. Biologists working in

Pfizer's Molecular Sciences research area emphasized the implications of the science for the company leadership. Time and resources were being spent developing drugs that didn't work well in a significant portion of users or, worse, had a harmful, unanticipated side effect. The resulting delay or even lack of approval from federal regulators could cost the company millions. The Pfizer scientists advised including specific genetic information from the very beginning of testing. If a particular compound, for example, was known to work through a particular metabolic pathway, then early trials of the drug could include a distribution of research participants with the different genetic variants that shaped that pathway. Or, if a particular compound was known to react negatively to a particular cytochrome P450 profile, then genetic testing could be included early in trials to weed out potential participants with that profile so that their poor response wouldn't skew the results.

The commercial and sales side of Pfizer, on the other hand, saw real economic challenges with pharmacogenomics. Drugs like Zithromax, Zoloft, and Zyrtec were huge moneymakers for Pfizer precisely because they could be prescribed widely and sold relatively inexpensively. Carving up patient populations based on which genetic profile people carried meant a new drug could be sold only to certain patients, a form of market segmentation, which necessitated selling the drugs at a higher price point. Higher-priced drugs, in turn, were harder to sell.

That financial concern was joined by a logistical complication for drugs developed with pharmacogenomics. A one-pill-fits-all drug, by definition, could be offered to everyone, so there was no need to worry about who could or couldn't take it. But if genetic differences between patients impacted whether some drug was appropriate, then that meant a diagnostic test had to be developed right alongside the drug in order to be sure that only the right people with the right biological profile got the drug. Diagnostics must also be developed and approved, so that meant the pharmacogenomic process required coordinating and rolling out a drug and a diagnostic, no small task when getting a drug approved by itself was very arduous. It also meant, even if the drug and diagnostic were both approved, convincing physicians and patients to accept this new approach to prescribing pharmaceuticals.

The scientific voices at Pfizer eventually won out. It was indeed hard to sell a drug that cost more, but it was also hard to sell a purported one-pill-fits-all drug that didn't in fact work for everyone. Moreover, there was an industry-wide sense in the 1990s that a genomic revolution was afoot, and a company had to either get on board or get left behind.

Pfizer, in 1997, added a Pharmacogenomics Section to its Molecular Sciences unit and started hiring geneticists focused exclusively on what Arno Motulsky described forty years earlier as "genetically conditioned drug reactions." In 1998 and 1999, it helped create and fund the SNP Consortium, the industry-wide collaboration among pharmaceutical companies and academic institutions to collectively invest in identifying single nucleotide polymorphisms in the human genome. Throughout the early years of the twenty-first century, it formed business agreements with dozens of smaller biotech companies that provided Pfizer with genetic services like screening potential participants in drug trials and helping to create a diagnostic that went along with a drug. So Pfizer was naturally receptive, in 2005, when Francis Collins reached out with an invitation to work together on a public-private partnership that used its industry dollars to shape genetic research conducted at the NIH.

. . .

In early 2005 neither the National Children's Study nor Collins's American Gene Environment Study (AGES) had financial support sufficient to launch a nationwide project. Both groups were searching for the dollars necessary to become a working reality. Collins figured, ever since the completion of the Human Genome Project, that pharmaceutical companies were a live option for his genetic work, and it was time to start exploring that route. In March, Collins traveled to Silicon Valley, where the offices of Perlegen Sciences sat adjacent to the picturesque Shoreline Golf Links and looked out over the San Francisco Bay. Perlegen, a spin-off of the large biotechnology company Affymetrix, was one of the premier firms with which Pfizer developed a business relationship in the early years of the twenty-first century; the smaller company provided Pfizer with genotyping services useful for drug development and testing. Biologists from both Pfizer and Perlegen greeted the head of the Genome Institute upon his arrival. Collins's proposal to them was for the three entities—the NIH, Pfizer, and Perlegen—to team up on a big-budget project managed by the Foundation for the NIH, that organization created by Congress in 1990 to facilitate such public-private partnerships. The goal was to conduct studies that identified genes associated with twenty common diseases, which would then pave the way for pharmaceutical interventions. Collins offered the private companies access to dozens of existing cohorts of research participants from studies spread out across the NIH, made up of people with all sorts of different diseases and disorders including osteoarthritis, major depression, and diabetes. In exchange, Collins asked Perlegen to use its proprietary genotyping technologies to perform the genome-wide association studies on the thousands of samples from the NIH cohorts. And Collins wanted Pfizer to pay for the endeavor, at a cost of well over $100 million, a big ask. Pfizer would get no exclusive access to the data, which would have to be deposited in a publicly available database. The Pfizer scientists agreed to take the proposal back to company leadership.

While waiting for Pfizer's response, Collins pursued a second line of support. In January, the U.S. Senate confirmed a new secretary of health and human services, Michael Leavitt, to oversee all things health related in the federal government, including the NIH, the CDC, and the EPA. Leavitt, before rising to the level of White House cabinet secretary, was a giant in Utah politics. A descendant of Mormon pioneers who settled in the southern region of the state in the nineteenth century, the savvy businessman was only the second governor of Utah to be elected to a third term. He didn't, however, finish that third term, because President Bush nominated him to lead the EPA in 2003. After serving only a year at the environmental agency, however, Leavitt was promoted by the president to lead all of Health and Human Services in his second term, where one of Leavitt's priorities was facilitating more interactions between the federal government and private industry. The Latter-day Saints of Utah have historically placed a high value on both large families and genealogy, which makes Utah pedigrees a magnet for human geneticists. After Leavitt was confirmed, Collins began communicating with several geneticists in Utah, figuring out what sort of genetic work the Genome Institute could bring to the new secretary's home state.

In mid-March, Collins crafted what he referred to as the "Leavitt-friendly proposal." First, the NIH would support genotyping work on the large Latter-day Saint families in Utah, which would entice the former governor. The NIH, Pfizer, and Perlegen would also strike their deal to perform genome-wide association studies on the twenty common diseases. Then, hopefully, Leavitt would step up and fund the multibillion-dollar study. Utah would have nothing to do with the Pfizer-Perlegen deal, and it would have little at all to do with AGES, but the thought was that the initial focus on that state would be enough to hook Leavitt.

The pieces of this plan started to fall into place by summer. In June, the Pfizer and Perlegen scientists traveled to NIH to work out in greater detail what the potential collaboration would look like. Pfizer was warming to the idea. Teaming up with the NIH gave Pfizer access to patient populations with all sorts of diseases that the pharmaceutical company hoped to start targeting for drug development. Beyond that, it liked the optics of working alongside the prestigious federal agency. It was true that the data from the genome-wide association studies would have to be made public, but the Pfizer scientists figured they were so far ahead of other drug companies in terms of pharmacogenomic investment and expertise that they'd be in the strongest industry position to take advantage of the data.

In July, Zerhouni, Collins, and David Schwartz traveled to make the Leavitt-friendly pitch to the Department of Health and Human Services. Schwartz was the new director of the National Institute of Environmental Health Sciences, another institute at the NIH. Since Collins and Duane Alexander parted ways in 2004, Collins sought a new partner to represent the environmental side of things, and Schwartz's role as leader of the NIH's institute devoted to environmental health made sense. Collins approached Schwartz just weeks after he arrived on the job and asked him to team up on AGES. Schwartz agreed but warned Collins that the *E* in his AGES was seriously underdeveloped. The Genome team's reliance on self-reported questionnaire data was not acceptable. They needed to actually measure the environment. Schwartz was a proponent of developing "labs on a chip"—sensors that would hang off somebody's belt buckle or key chain and acquire data about the environments through which the wearer moved. The problem, he admitted, was that the technology wasn't there yet. Mass-producing and implementing such devices was more aspiration than reality in 2005.

Zerhouni, Collins, and Schwartz's proposal for the secretary of health and human services was a series of projects designed to build up to the American Gene Environment Study: the genetic study of the Utah population; the technology development on Schwartz's side; polling of public perceptions regarding participating in a nationwide study; a small-scale version of AGES with just 30,000 Americans; and finally Collins's full study with 500,000 participants. There was no inclusion in their joint presentation of the one NIH nationwide study already in the works—the National Children's Study. The NIH delegation, as a result, was invited to come back in August alongside Alexander to consider what a coordinated effort between the two teams might look like.

When the Genome team arrived with a sudden invitation to rapidly prepare for an audience with Health and Human Services, it complicated an already hectic schedule for the National Children's Study. After Alexander decided back in 2004 to go with the national probability sample rather than a convenience sample, Child Health worked with the CDC's National Center for Health Statistics to carve up the United States into representative units based on census data. Just over a hundred sampling units were randomly selected, marking the study locations; then, from those study locations, census blocks (essentially neighborhoods) were randomly chosen again. The plan was for recruiters to knock on every door in those randomly selected neighborhoods to identify women who either were in the early stages of pregnancy or were between the ages of eighteen and forty-four and so might become pregnant. The study locations and recruitment at them were going to roll out in waves over several years. At the front of the line were seven sites dubbed the "Vanguard Centers": Queens, New York; Orange County, California; Montgomery County, Pennsylvania; Waukesha County, Wisconsin; and Salt Lake County, Utah, were all in urban settings; Duplin County, North Carolina, and the combination of Brookings County, South Dakota, with Lincoln/

Pipestone/Yellow Medicine Counties, Minnesota, were quite rural. Scientists at universities affiliated with hospitals that served those communities applied to take charge of the study locations, following the example set by the Collaborative Perinatal Project. Landrigan, by then leading the children's health program at Mount Sinai Medical Center in New York City, wanted to serve as principal investigator for Queens, the most diverse site. In the summer of 2005 the National Children's Study personnel were receiving applications that were more than a thousand pages long and making site visits to see if the institutions were equipped for the enormous undertaking.

Selecting the Vanguards also revealed a new challenge for Alexander. The budget for the National Children's Study was growing. When scientists in the program office of the study took stock of all the tests, measures, and experiments that the working groups were proposing, they realized that the total direct costs were more than initially budgeted at roughly $3 billion. What's more, the initial budget included no indirect costs. When the federal government teams up with outside collaborators, like academic institutions, the contracts include administrative overhead that covers all the expenses associated with actually running the project. Those administrative costs can be 50 percent on top of the direct costs of the project. The anticipated costs of inflation also needed to start getting built in since the project was expected to last two decades.

Alexander's institute budget had to be approved every year by budget officers of the whole NIH who in turn passed the NIH budget on to budget officers in Health and Human Services, so Alexander requested permission to start reporting a National Children's Study budget that included a new, higher cost estimate. To his surprise and disappointment, the budget personnel told Alexander that he could not update the study's total projected cost. He had to stick with the number he'd already reported. That put him and the study that his institute hosted in an uncomfortable position. He knew that the grand vision being conceived by all the National Children's Study affiliates, which he encouraged, didn't match what could actually be afforded in the current projected cost. The budget directive, however, constrained what he could publicly say about the price tag. If the NIH leadership would not let Child Health report a higher number, then that also meant Alexander could not tell the media that it might cost more or even tell Congress that it might cost more. Three billion dollars was the only number that the NIH would permit, and so that had to be the official number for anybody who asked about the estimated costs of the study. Alexander, as a result, would have to wait until the next budget cycle, the following year, to try asking again.

In the meantime, there were two solutions for operating under the budget directive that Alexander received. One option was trimming the whole study down after the Vanguard Centers conducted the pilot testing. The purpose of the Vanguards, after all, was in part to see what worked in the children's homes, what didn't, what could be afforded, and what couldn't. So there would be places to set limits. Alexander conferred with his Child Health team, and they confirmed that they could run a pared-down version of the study that included all the costs at the $3 billion total. It might not be the National Children's Study that everyone had been imagining, but it would be a National Children's Study that met all the basic criteria as laid out in the Children's Health Act. The other option was tapping into some new sources of support. That might be organizations independent of the federal government, or it might be additional enthusiasm for the study elsewhere in the federal government. Ever the optimist, Alexander trusted a solution would eventually arise.

The opportunity to present the National Children's Study to Secretary Leavitt was, for Alexander, an exciting development because it dangled the prospect of federal support. It was also worrisome because it was being orchestrated by Collins and came out of the blue. The National Children's Study, unlike Collins's endeavor, had few routes for securing financial support from private industry. Companies that produced environmental toxins had little interest in financing research that specified just how dangerous those were for children. The pharmaceutical industry, for its part, had no incentive to invest in uncovering environmental health threats unrelated to drug sales. Nonprofits were a viable option for funding—a nonprofit committed to children with asthma, for instance, might make an investment in an aspect of the study devoted to investigating air quality in children's homes—but they could hardly fund the entire thing. As a result, it came down to either Congress or the White House making a major commitment, and the opportunity to present to Health and Human Services opened the door to the administration track. The Child Health team spent three frantic weeks in August communicating with the Genome team, brainstorming ways to share resources such as recruitment sites, data repositories, and ethical oversight of their separate projects.

When Alexander met with the secretary alongside Zerhouni, Collins, and Schwartz, it became clear that he was joining a conversation that had started without him. They had met previously when Alexander wasn't present, and it wasn't at all obvious how the National Children's Study fit into the vision that was unfolding. That concern was compounded when communication between Genome and Child Health dropped off again as soon as the co-presentation to Leavitt's office was completed. It was as if a box had been checked, and Alexander was once again out of the loop. Two months later, the secretary made a special trip to attend the American Society of Human Genetics meeting, where he joined Collins in hailing the value of federal investment in personalized medicine. By the fall of 2005, it was clear to Child Health that they were getting politically outflanked, that the "backup plan" Collins proposed in 2004 was now threatening the existence of the National Children's Study. Fortunately, Alexander was about to get assistance from the expanding community of study affiliates.

Child Health announced the institutions and scientists who were going to run the Vanguard Centers that fall. In November, Landrigan, representing Queens, and the other leaders of the newly selected recruitment sites gathered at the NIH to meet one another as well as the study's leadership. The official agenda revolved around logistical items for the Vanguards—reaching out to local community leaders, strategies for advertising the study in the neighborhoods, hiring personnel to do the recruiting. During coffee breaks, though, conversations shifted to rumors of the National Children's Study getting shoved aside in favor of Collins's project.

After the meeting Landrigan had dinner at one of Bethesda's Italian restaurants with Edward Clark, chair of pediatrics at the University of Utah, who was just named head of the Vanguard Center in Salt Lake County, and Alan Fleischman, a neonatologist and pediatric ethicist who had recently joined the study's program office part-time as an ethics adviser and also chair of the Federal Advisory Committee. Surrounded by the smell of marinara sauce and garlic bread, the three took stock of the gathering they'd just sat through. All three were renowned pediatricians and politically astute administrators at their prime. They agreed that the study was in danger, and the political threat required a political response. While brainstorming strategies, Landrigan recalled a recent awards ceremony hosted at Mount Sinai, where John Porter, the former congressman from Illinois and lifelong supporter of federal investment in biomedical research, was honored. Porter left office in 2001 after serving two decades; in 2005 he was both a senior partner at the prestigious Hogan & Hartson law firm in Washington, D.C., and a member of the board of directors at the Foundation for the NIH. Landrigan had a chance to speak with Porter at the ceremony and knew he was passionate about children's health research. Clark liked the suggestion, but before committing, he wanted to meet Porter in person to feel him out.

The two Vanguard directors, the next day, hopped on the D.C. Metro's Red Line and rode it downtown. The train ride wasn't nearly so long as Landrigan's flight from El Paso to Austin thirty years earlier, but the goal was very much the same—to find a political lever that allowed for removing an obstacle in the way of important science. When they reached Porter's office, Landrigan and Clark explained the situation to the Washington insider. They told Porter that there was no immediate need to retain him, but if political support were to suddenly shift away from the National Children's Study, that would change things. Clark liked what he saw, and Porter, a supporter of the Children's Health Act when he was in the House, appreciated the opportunity to advocate for it. The three agreed to stay in touch as the early months of 2006 unfolded, when the federal budgetary commitments of the next fiscal year would be revealed.

While Landrigan and Clark were formulating a tentative partnership with Porter as a lobbyist, Collins's partnership with Pfizer and Perlegen came together. Pfizer was not willing to pay for the entire set of twenty genome-wide association studies, but it would purchase the genotyping capacity at Perlegen to undertake the first five, each involving about a thousand cases and a thousand controls and looking at 375,000 different SNPs in each participant's genome. Pfizer was also going to contribute an additional $5 million to the Foundation for the NIH, dedicated to organizing and administering the public-private collaboration. In exchange, Pfizer got to place its own scientists on the governing committees, which ensured that the company contributed to the decisions about which diseases would be prioritized and which NIH cohorts would be utilized. Collins also managed to bring Affymetrix into the mix, one more source of industry funding; it agreed to take responsibility for two more genome-wide association studies. All told, private companies provided roughly $26 million to the investment in gene hunting called the Genetic Association Information Network.

Leavitt also got behind Collins's plan. The secretary wasn't willing to commit the hundreds of millions of dollars each year it would have taken to launch AGES, but he was willing to commit $40 million annually for the remaining years of the Bush administration, and that allowed Collins and Schwartz to launch what they called the Genes, Environment, and Health Initiative, which would provide $16 million to Schwartz for his tech development efforts and $24 million to Collins to conduct even more genome-wide association studies. In the end, between the two initiatives, Collins received $50 million (roughly half from private companies, half from taxpayers) to hunt for genes associated with common diseases. With both industry and the federal government opting for Collins's approach, the question remained: What to do about the National Children's Study?

In early February 2006, the White House's proposed budget for fiscal year (FY) 2007 was released. No funds were included for the National Children's Study. This was not a surprise; the study had been moving along without federal financial support for several years. But the next lines were a shock: "The [National Children's Study] planning activities that are ongoing under contract in FY2006 will be brought to a close by the end of the year. There are no plans for the NIH to continue to pursue the full-scale study in FY2007." Collins admitted back in 2004 that his machinations to work around the National Children's Study would be "a bit messy" unless Zerhouni got rid of Alexander's project, a prospect he thought unlikely at the time. Now the FY2007 budget, with its call for terminating the study, did exactly that.

The scientists at Child Health scrambled when they received word of the administration budget. Fleischman stepped back from his desk, packed up his briefcase, walked out of the building, and off the NIH campus. He was a part-time employee of Child Health and so a part-time federal employee. While at the NIH, at his federal desk, and in front of his federal computer, there was nothing he could do about a presidential budget that called for eliminating the project on which he was employed. But once he left the physical premises, he could switch roles and act. Fleischman surreptitiously obtained a list of all the congressmen and their health staffers as well as a list of the top health reporters in the country. He spent the ensuing hours contacting them one by one, explaining what a tragedy it would be if America's first national study of children in decades was suddenly eliminated.

Full-time federal employees at Child Health, like Alexander and Scheidt, could not follow in Fleischman's footsteps, but they could contact those who had that ability. The Salt Lake County principal investigator, Clark, was sitting in an airport terminal in Newark when he received Scheidt's anxious call. Clark, in turn, conveyed the message to Landrigan. They agreed it was time to bring in Porter.

As the National Children's Study scientists were circling their wagons, Zerhouni, Collins, and Schwartz took the stage at the National Press Club. The club sits atop the National Press Building just two blocks from the White House, a swanky gathering place for journalists and news producers that also serves as a desirable venue for big media events, such as presidential candidacy declarations and major policy announcements. The news that day involved the NIH's major health initiatives for FY2007—the Genetic Association Information Network, and the Genes, Environment, and Health Initiative. Zerhouni, Collins, and Schwartz were joined by Martin Mackay, a vice president from Pfizer, and, in an ironic twist, by John Porter. He was on the stage representing the Foundation for the NIH, and unbeknownst to the others he was already preparing to make good on his pledge to save the National Children's Study.

The reporters who packed into the Murrow Room, where the event took place, were surrounded by shades of blue—blue rug, blue table wraps, blue chairs, blue wall, blue National Press Club logo. Microphones were set up for members of the press in the room to ask ques-

From left to right: Francis Collins, Elias Zerhouni, and David Schwartz attend the February 2006 National Press Club event; John Porter is barely visible on the far right. *NIH Record*

tions, while an audio system permitted journalists from afar to phone in, listen, and pose queries of their own. Cameras flashed as the well-dressed figures took their turns at the lectern. Zerhouni emphasized that public-private partnerships were a priority for his administration of the NIH. Collins predicted the programs would "fuel this era of preventive, personalized, preemptive medicine." Mackay situated the company's involvement in its broader embrace of pharmacogenomics and its search for new pharmaceutical cures. Schwartz wore a teal-shaded prototype that dangled off his lapel and asked the reporters to imagine a future where that sensor device could measure dozens of chemicals in the environment and tell him and his doctor whether he'd been exposed to something that put his health at risk.

When reporters were given the opportunity to engage the speakers, the exchanges initially remained light. The scientists at the front of the room complimented the questions posed. They cracked jokes about dysfunction in Congress and received polite chuckles. Then a reporter brought up a topic that shifted the mood in the room. "A lot of the focus from this research is very similar to the focus that was part of the National Children's Study, in terms of diabetes, autism, asthma, and the interaction between environment and health," he pointed out. "But this program has a $40 million boost, and that program lost funding, and so I'm just wondering if this represents a shift in research focus from the NIH." Collins, seated, glanced down momentarily and then straightened his posture, turning his attention to Zerhouni, who was standing at the lectern. The NIH director, shifting his weight back and forth from right foot to left but maintaining a comfortable grin, replied that the two decisions were unrelated. "The juxtaposition is a correlation, not a causation." Had Zerhouni left his reply there, the matter might have been dropped. Instead, he kept talking. The United States could one day attempt a large, longitudinal study like the National Children's Study, Zerhouni said, but these genome-wide association studies had to come first in order to set the foundation for that more ambitious effort. "I really would like to dispel the notion that this was a decision of one against the other." In his protestations, Zerhouni managed to contradict himself over the course of just sixty seconds. He wanted the reporter to believe that the decision to terminate the National Children's Study had nothing to do with launching the new initiatives and simultaneously that something like the National Children's Study had to wait for the results of the new initiatives.

Zerhouni, until that point, had been controlling the reporter-speaker interactions, directing a question about environmental measures to Schwartz, one about drug applications to Mackay. But now Porter, sitting just to Zerhouni's left, jumped in. "I just want to say, wearing another hat, that there are many of us who believe that the National Children's Study is of tremendous importance to the future of health in this country and want to be a strong supporter of that study and will be a strong supporter of that study and not accept the decision of the White House to zero it out in the current budget context." Porter then came as close as he could to saying publicly that he'd be lobbying for the study, concluding, "I'm very hopeful that this study can continue and will be working very hard to make that happen."

There was no immediate reaction to Porter's interjection from the reporters present, and Zerhouni appeared pleased to move on to other questions. But just when it seemed as if the event were wrapping up, another journalist picked up on the inconsistency in Zerhouni's answer about the National Children's Study. "I just wanted to make sure that I understood fully," she began. "Are you saying that you personally decided that the science was not right to continue with that study at this moment . . . ?"

Zerhouni's smile and pleasant demeanor cracked; now tense and defensive, he threw his hands up and interrupted. "No, I didn't say that at all. No. I don't know where . . ." He trailed off. "The implication was that there was a zero-sum game of this initiative sort of making a decision for the other. I'd like to dispel that. There is absolutely, uh, no connection between the two. I think a decision was made that, uh, basically relates to budgetary issues." Collins slowly nodded his head in agreement. But when the reporter tried to ask for more specifics, Zerhouni interrupted her once more, conveying his belief that the genome-wide association studies had to proceed before a nationwide study could be conducted. Moreover, Zerhouni wanted the reporter to know that there were different ways to design such a study. Echoing one of Collins's talking points about the National Children's Study, Zerhouni said valuable health information would take eighty years to acquire because the scientists would have to wait for the children to grow up. The NIH was considering another project that didn't have that limitation—Collins's American Gene Environment Study, which included participants of all ages. "I think I want to dispel the notion that the one took from the other. It really isn't, at least from the NIH standpoint, a consideration." The NIH director glanced awkwardly over at Porter and then thanked the reporters for coming.

The months that followed were chaotic for the National Children's Study. On the one hand, preparations needed to be made to shut the operation down. Termination meant disbanding the working groups and advisory committees, closing down the program office for the study at the institute, ending contracts with the new Vanguard Centers. A massive organization had coalesced around the National Children's Study over the previous six years, and the official plan had to be that everything was being brought to an end.

On the other hand, preparations unofficially needed to be made to launch the project. If the lobbying and media outreach efforts paid off, leading Congress to step up and provide funding, the study needed to be ready to hit the ground running. That meant having the Vanguard Centers equipped to advertise the study to their communities before actually going door-to-door, translating all the working groups' recommendations into a realistic study protocol, and getting everyone affiliated with the effort clear on how the data was going to be collected and stored.

Fortunately, by 2006, Scheidt did not have to navigate the frantic period alone in the program office. Closest to Scheidt was Sarah Keim. In 2000, she was participating in the prestigious Presidential Management Intern Program, which gave recent graduates interested in public service the opportunity to gain leadership and management training while interning with federal agencies. Keim wanted to work at the intersection of science and policy, so when she heard about the National Children's Study, she arranged to do her internship with Scheidt. For the early years of the project, it was just the two of them getting it up and running, which gave them plenty of time to get to know each other. Scheidt, decades Keim's senior, mentored the young scientist, assisting her with earning a PhD in epidemiology while she was there. Keim also occasionally joined Scheidt's family for the Wednesday evening boat races, although, by her telling, it was only to serve as ballast.

The program office, however, grew to more than a dozen staff by 2006, and Keim rose to serve as a deputy director in it. That put Keim in close contact with the Vanguard Center directors, as well as the various advisory bodies and working groups. She was also intimately aware of the budget directive that Alexander received in 2005, prohibiting him from adjusting the total projected cost of the study, and she was often put in the awkward position of explaining why the National Children's Study price tag looked the way it did as it moved up through the various budget offices. When a budget officer in Child Health, for example, asked in 2006 why inflation wasn't included, Keim replied, "To date, we haven't been incorporating inflation into the numbers in order to maintain the feeling that the total cost of the Study is within some realm of fund-ability." Since the budget directive constrained them to the number they'd been reporting previously, Keim pointed out that the NIH leadership "may react negatively" if the program office suddenly supplied them with a much higher number.

Alexander hoped to go back to the NIH leadership in 2006 and get out from under the budget directive he'd received the previous year by asking once again for permission to report the new projected budget of the National Children's Study, with all the anticipated direct costs, indirect costs, and inflation incorporated. But the shocking announcement in February upended that plan. The cost estimate, Alexander was told, could not be changed. The administration's plan was to shut the study down, not expand investment in it. The official message from Child Health to members of Congress and journalists throughout 2006 couldn't be about the need for more money; it had to be about bringing it to an end.

Fighting the prospect of termination, as a result, fell to others. That's where the dinner among Landrigan, Clark, and Fleischman months earlier paid off. Clark took charge of serving as "banker," collecting money from the Vanguard Center directors. Fleischman informed them that the funds couldn't be tied to the federal government in any way, because federal dollars may not be used to pay for lobbying the federal government. He even helped them craft a script they could use to get the resources from their upper administrators: the academic institutions might need to invest tens of thousands of dollars now to save the National Children's Study, but that would be dwarfed by the millions of dollars that would come back to the institutions when the Vanguard Center pilot testing began. Clark then turned the collected funds over to Hogan & Hartson to retain Porter, who worked for a fraction of his standard billing rate.

Porter, only a couple of years removed from serving as influential chairman of the appropriations subcommittee responsible for allocating taxpayer dollars to Health and Human Services, remained well connected with his former colleagues in Congress. He contacted current members in the House and Senate—the ones who would decide where the nation's federal dollars went in 2007. His case to them was simple and personal: we authorized the National Children's Study back in 2000 when we passed the Children's Health Act, so it was time to step up and save it.

Zerhouni, each spring, appeared before the House and Senate appropriations subcommittees to provide testimony about his agency's budget and initiatives for the coming year. The occasions, held in ornate rooms of Congress where elected officials stare down at their guests from leather chairs and desks of dark wood and marble, tend to be cordial affairs. The NIH historically has a very good relationship with Congress; politicians like to publicly show their support for biomedical research, and time with the NIH director presents such an opportunity. In the spring of 2006, however, Zerhouni faced the repercussions of Porter's advocacy. When asked to explain the termination of the National Children's Study to representatives in the House, Zerhouni said that with the big investment that the federal government was making in the Genes, Environment, and Health Initiative, alongside the private industry support for the Genetic Association Information Network, the timing just wasn't right for the National Children's Study. Rosa DeLauro, a Democrat from Connecticut, wasn't convinced; the Vanguard Centers were already rolling out, she said, so shutting the whole thing down now "just doesn't make sense."

Before the Senate, Zerhouni fared even worse. He justified the decision to the senators as "just a matter of budget priorities," but that didn't sit well with the committee members who supported it in the Children's Health Act. Tom Harkin, a Democrat from Iowa and one of Porter's first contacts, looked over at Senator Arlen Specter from Pennsylvania, who was chairing the committee, and angrily declared, "You and I and others on this committee had planned for this children's study. It was passed in 2000. A lot of planning went into this and forethought went into it to set up this long-term study, and I just cannot believe that we are just going to just stop it at this point in time." The president's budget was one thing, but the federal budget was ultimately controlled by Congress, and Senator Harkin wanted Congress to mandate funding it.

Fleischman's media blitz also proved effective. Analyses performed for the study leadership indicated that there had been more than 150 media placements about the study in the months following the president's budget announcement. That coverage was extremely sympathetic to their plight. An editorial in *Scientific American* argued that the significant dollars it would cost to fund the National Children's Study were just a fraction of the billions devoted to health-care costs associated with things like autism and asthma. If the study can contribute even a small amount to reducing the incidence of those conditions, the editors pointed out, "it would be a fool's economy to squander the money and effort already spent by killing this effort in the cradle," read the piece titled "Don't Rob the Cradle," and concluded by asking members of the House and Senate, "Do you know a bargain when you see it?"

As 2006 unfolded, support for the study gained momentum. In November, Democrats took back both the House and the Senate. Affiliates of the study welcomed the political change, anticipating the progressive party in Washington to be more sympathetic to their focus on children's health and the environment. Sure enough, on Valentine's Day 2007 the news that Child Health had been waiting for finally came. In the continuing resolution budget for 2007, Congress allocated $69 million for the study. It was only a one-year commitment that would need to be reviewed again the following year. Still, the study was now a funded mandate, and that meant its future was far more promising than it had been at any point in its history. Collins had his money from Pfizer and Leavitt, but Alexander had the money from Congress. Back at Child Health, Alexander's staff were overjoyed when the funding news came in. Collins's attempt to turn their merger into an acquisition had been deterred.

4

From Artificial Nose to Genomics Juggernaut

IN THE WORLD of genomic technology companies, Illumina is without peers. It creates the machines, the chemical reagents, the sensors, and the services that extract information about genetic sequence, genetic variation, and genetic function from biological samples. That might be spit in a test tube that somebody mailed to a personal genomics company, blood in a vial stored for years in a government laboratory, or a specimen provided by someone participating in research conducted at a university. More than 90 percent of the world's DNA sequence information is generated with Illumina products. The company, whose mission statement is to "improve human health by unlocking the power of the genome," has operations across the globe, from its headquarters in San Diego to France, Germany, Brazil, China, Japan, South Korea, and Australia. It is valued at more than $50 billion. Illumina holds roughly eight hundred patents, with hundreds more pending. The science reporter Sarah Zhang pointed out that you could call Illumina "the Google of genetic testing," but the comparison would understate the company's dominance. Illumina has nearly eight thousand employees, but it started out as a conversation among three people—who met to chat about artificial noses.

By 1997, Larry Bock was a known entity and proven success in the world of life science venture capital firms. His specialty was incorporating innovations from chemistry into the health sector. Just a year earlier, his company, Pharmacopeia, went public, utilizing combinatorial chemistry to develop pharmaceutical compounds. Now he was looking to get in on the artificial nose business. The concept of an artificial or electronic nose itself really wasn't new. The basic idea was to model a sensor device on the olfactory system. Just as your nose uses chemical information to distinguish the smell of a cup of coffee from freshly cut grass, the goal of the artificial nose was to create something that could distinguish exposures based on the unique chemical patterns they produced. The applications of such a technology were limitless: specifying when a cut of meat had begun to spoil, pinpointing the precise time to harvest cabernet grapes, dialing in the exact recipe for mass-producing a popular perfume, detecting explosive materials, identifying a toxin in drinking water. Bock first approached a team of renowned chemists at Caltech, but he learned to his disappointment that they'd already committed to working with a different set of investors. That led him and his associate, John Stuelpnagel, to have breakfast with David Walt, an analytical chemist at Tufts University who'd recently developed a new technology that could multiply by orders of magnitude the number of data points that a single sensor device could collect.

Walt was born and raised in Detroit and drawn to the outdoors. He went to college at the University of Michigan, where he fell in love with chemistry because it allowed one to manipulate nature at a fundamental level, producing chemical phenomena and products that might never have been seen before on Earth. Chemistry took Walt to graduate school in New York City, where he earned a PhD at Stony Brook, and then up to Massachusetts as a young assistant professor at Tufts. Walt quickly made a name for himself in the domain of chemical sensors by taking advantage of the growing science of fiber optics. Fibers made of glass or plastic can transmit light with very little signal loss. That, combined with their flexibility and small size, makes them wonderful tools to communicate information; used for research purposes, "information" equals data. In an early example of this, Walt developed a method for attaching fluorescent dye to the end of optical fibers, which chemically responded when exposed to substances like benzene and gasoline. He then took the device to Pease Air Force Base, on New Hampshire's Atlantic coast. The base had recently been declared a Superfund site by the U.S. Environmental Protection Agency because jet fuel leeched into the surrounding soil and water, leading to growing concerns about the health impact on the soldiers stationed there. Walt, funded through the EPA, tested his new device. By dipping the dye-coated ends of the illuminated optical fibers into several different wells, he was able to confirm the different levels of jet fuel contamination in the different locations.

Walt's entry into the science of artificial noses was an extension of the Pease work, but rather than obtaining data from a single fluorescent-dye-coated fiber, he obtained data from a bundle of them. That multiplication of sensors was the essence of the artificial nose. The design of the first working electronic nose was published by Krishna Persaud and George Dodd in 1982. It was an exercise in biomimetics—modeling an engineered device on a system found in nature. When you sniff air in through your nose, molecules from the surrounding environment rush into your nasal cavity. The olfactory epithelium is a patch of tissue high up in that cavity almost adjacent to your brain, where millions of olfactory cells reside. Most humans can detect tens of thousands of smells, but you don't have tens of thousands of different olfactory cells—one cell for coffee detection and another for grass detection. Rather, you have a couple hundred different types of olfactory cells. A smell, as a result, is not the detection of a particular molecule by a particular olfactory cell but the detection of a pattern of molecules across a whole set of olfactory cells. Your brain then does the work of distinguishing the coffee pattern from the grass pattern. An artificial nose replicates this process with three key components: the delivery system (equivalent to your sniff), the sensor system (equivalent to your olfactory epithelium), and the pattern recognition software (equivalent to your brain).

Walt and the graduate student Todd Dickinson joined forces with neuroscientists at Tufts to create an artificial nose that utilized optical fibers as the foundation for the sensor system. They fabricated a bundle containing nineteen separate fibers, each with the fluorescent dye on the end. In that sense, it was similar to the device Walt inserted into the wells at Pease. This time, though, the dye on each tip was secured on a unique polymer matrix that had a specific polarity, flexibility, and pore size, which meant each fiber was responsive in its own way. This meant data could be obtained from all the sensors at the same time and combined to look for patterns. The fluorescent pattern of benzene, it turned out, was different from the fluorescent pattern for propyl alcohol. They'd created a working model of an artificial nose in the lab, and because the version utilized an entirely new application of optical fibers, they started calling it their "optical nose," so as to distinguish its design from the other, existing electronic noses.

As innovative as Walt's optical nose was, it did have a very practical limitation. Fabricating the device was time consuming because each individual fiber needed to be equipped with its own individual matrix/dye tip. If you wanted more sensors, then the fabrication process took that much longer because additional sensors needed their own tailored tips. A radical answer to that problem presented itself when Walt's graduate students briefly dipped the tips of the fiber-optic bundles into hydrofluoric acid, a highly corrosive solution, which chemically etched shallow wells only a couple of micrometers deep in the tips. The result was something that under an atomic force microscope looked like a sprawling egg crate. That, for Walt, lent itself to a follow-up question: Could they reliably put something in those indentations at the end of each fiber? Microbeads just microns across were commercially available at the time. The beads were small enough to fit in the divots; moreover, they could be coated in substances that remained affixed to the beads even after they settled. When they looked at the atomic force microscope images of that subsequent effort, it was as if tiny eggs had found their ways into each spot in the egg crate. They were staring at an array of microwells on the tip of a fiber-optic bundle where microbeads had come to rest, what they described as a "bead array."

A fiber-optic bundle only one millimeter in diameter could house tens of thousands of fibers. Dip the tip in hydrofluoric acid, and you would have a fiber-optic bundle with tens of thousands of microwells just minutes later. Mix up a batch of microbeads with the desired chemicals affixed, like fluorescent dyes, and you would have a bead array with tens of thousands of spheres resting on the tip of the fiber-optic bundle. In just a couple of years, Walt and his graduate students had gone from fashioning an artificial nose with nineteen sensors, each of which had to be individually constructed, to one with tens of thousands of sensors that were mass producible in a fraction of the time. Whether you wanted a device with 100, 1,000, or 100,000 sensors, the exact same process produced the desired result depending on how many fibers were in the bundle and what you affixed to the beads. That enormous increase in the number of sensors on such a minuscule scale, the ease with which they could be fashioned in a replicable manner, and the fact that the essential pieces of the device—the fiber-optic bundle, the microbeads, the hydrofluoric acid—were all readily available were what made the bead array so promising.

Bock was excited about the prospect of breaking into the artificial nose market with Walt's bead array innovation, but the artificial nose was not the only potential application of the technology. Walt, passionate about both basic science and entrepreneurship, saw that immediately. The bead array was a breakthrough in sensor technology; sensing chemicals in the environment was one natural application, but so too was sensing different sequences of DNA in biological samples. The microbeads could be coated in fluorescent dyes to sniff out the unique patterns that different compounds produced. They could also be coated in short strands of synthetic DNA called "oligos" (short for oligonucleotides). The presence or absence of some gene of interest, the tendency of some tissue sample to express certain genes, the location of a single nucleotide polymorphism—all of those things could be queried on a massive scale with bead arrays once the appropriate oligos were synthesized. The natural laws of chemistry did the rest when prepared samples were run across the arrays and the genetic material in the samples bonded to the complementary oligos attached to the microbeads.

The implications for genetics seemed just as exciting as the artificial nose applications, but Bock, Stuelpnagel, and Walt were not geneticists themselves. Bock and Walt were chemists by training; Stuelpnagel's background was in veterinary medicine. Mark Chee, who had been director of genetics research at Affymetrix, was invited to take a look at what Walt had. He was about as good a person as anyone to judge the potential for a new company to enter the genomic technology business. Affymetrix, at the time, was the eight-hundred-pound gorilla in that space; its GeneChip was the gold standard in the industry for determining what genes were being expressed in a sample. Any new company that wanted to get in the game needed to do what Affymetrix did better or carve out an original niche. Walt and his graduate students prepared to host Chee for a month in early 1998, allowing him to spend time in the lab, see how the microwells were created, treat the microbeads with various coatings, and examine the atomic force micrographs. After just three days, Chee told Walt they were sitting on something big, and Chee wanted in.

A company was incorporated in the spring of 1998, just months after Walt had first met with Bock and Stuelpnagel. The three of them, Chee, and the chemist Tony Czarnik, who served as the original chief scientific officer, were the co-founders. According to Czarnik, they set aside fears concerning conspiratorial associations with the secret society Illuminati and embraced the illuminated optical fibers at the core of the bead array technology with "Illumina."

The new company licensed that technology from Tufts for any application, but it quickly became clear to the group that they needed to act fast on the genetics side for two reasons. First, "personalized medicine" was then all the rage. Start-ups like Genaissance and drug giants like Pfizer were both drawn to the prospect of using pharmacogenomics to replace the one-drug-fits-all approach with the right drug, for the right patient, at the right time. To do that, they needed lots and lots of genetic data—on patients, on research participants, on biological samples sitting in cold storage across the globe. As the biomedical industry made this shift, the Illumina team wanted to be the go-to company for providing the genetic information that was intended to make personalized medicine a reality. Second, Affymetrix already had a head start. With Chee on board, though, they knew where Affymetrix's vulnerabilities lay and where the bead array could prove most competitive.

The Illumina scientists spent the next two years expanding and preparing to go public. Dickinson, Walt's graduate student, signed on as one of the first employees. Jay Flatley, who'd impressively led another biotech company, joined as CEO. The potential for Illumina's patented "BeadArray" technology to upend the genomics marketplace proved a valuable lure. An initial round of investors generated

$8 million in 1998. The following year, another round of financing raised more than $25 million. Illumina went public in the summer of 2000, and excitement only increased. The six million ILMN shares were initially priced at $9–$11, but by the day of the IPO they'd climbed to $16.

The thrill was short-lived. Illumina's IPO was among the last issued before the dot-com bubble of 2000 burst. Illumina was not an internet-based company, which bore the brunt of the financial collapse. Still, it was impacted by the financial uncertainty that stretched across the tech industry. The Illumina shares fluctuated radically, climbing to more than $22 months after the listing and then plummeting to less than $2 by 2002.

Illumina rode out the roller coaster. The leadership, scientists, and staff knew they had a genuinely innovative technology, and they delivered by rolling out an expanding lineup of devices with Walt's bead array at the heart of the system. Their BeadLab appeared in 2002, then the BeadChip in 2003, then the BeadStation just months later. It quickly became clear that Affymetrix had a real challenger on its hands. The bead array permitted such a high density of sensors, such scalability, and such flexibility in terms of what could be detected that it completely altered what geneticists could expect from the microarray industry. Chee, additionally, knew that Affymetrix was not investing as heavily in SNP detection, opting instead to prioritize other genomic applications. The early years of the twenty-first century were exactly when human geneticists began taking Neil Risch and Kathleen Merikangas's proposal for massive genome-wide association studies seriously, and those were premised on obtaining information about whether a sample had an A, a C, a G, or a T at hundreds of thousands of different locations. Illumina, as a result, positioned itself to lead the SNP detection enterprise, thereby cornering the genome-wide association study market. Moreover, when Illumina couldn't naturally do something itself, Flatley wisely acquired a company that could. Rather than developing the technology to mass-produce the oligos affixed to the microbeads, for example, Illumina acquired Spyder Instruments in 2000 and its proprietary tilted centrifugation technique to build Illumina's "Oligator Farm." As interest among geneticists expanded beyond information about

SNPs to information about the whole sequence of genomes, Illumina acquired Solexa in 2007 for its sequencing technology at a price of more than half a billion dollars.

During the first decade of its existence, Illumina also formed a symbiotic relationship with the genetic endeavors taking place at the NIH's Genome Institute. The first major example of this was the International HapMap Project, which launched in 2002. That effort, which took shape as the Human Genome Project was wrapping up, aimed to identify millions of SNPs across the human genome. While humans are more than 99 percent genetically similar, the goal was to home in on that less than 1 percent where variation existed and was thought to be associated with different health outcomes. Scientists obtained genetic information from about 270 individuals in the United States, Nigeria, Japan, and China, seeking to find locations where the genetic variation lay. The group announced their discovery of more than one million SNPs in 2005 and followed that up with another batch of three million SNPs in 2007. Illumina both benefited from and significantly contributed to the HapMap Project. Chee and Illumina received more than nine million taxpayer dollars to take the lead as one of the eleven centers, but that told only a part of the story. Of the other ten centers, half were relying on BeadArray machines that they'd purchased from Illumina. All said, most of the millions of SNPs identified through the International HapMap Project were isolated on a device with Walt's bead array in it.

Yet another example was the Genome Institute's Electronic Medical Records and Genomics (eMERGE) Network. It was housed in the Genome Institute's Office of Population Genomics, the same unit that hosted the Genetic Association Information Network with Pfizer and the Genes, Environment, and Health Initiative with the NIH's Environmental Institute. eMERGE rolled out in 2007 alongside those initiatives, and, like the other two, it represented a major player in conducting genome-wide association studies. But the key partners in it were the nation's massive hospital systems, such as the Mayo Clinic and Vanderbilt University, that had both robust electronic medical records and biobanks of samples from their patients. Francis Collins hoped to harness the power of data regarding health outcomes in the hospitals' electronic records and partner that with data regarding the genetic information of the patients from whom those electronic records were derived. Since Collins didn't have the funding necessary to create his nationwide American Gene Environment Study, initiatives like those in population genomics were all designed to lay the foundation for a future nationwide study. The eMERGE group looked for genes associated with traits ranging from cataracts to diabetes, and once again Illumina's technologies proved central to accumulating the genetic data about the patients.

Another company entered the genomics industry with a radical new business model around that same time. 23andMe, backed by Google's deep pockets, first offered its direct-to-consumer genetic testing services to customers in 2007. With a vial of spit and $1,000, someone could find out if they were at risk of age-related macular degeneration or to what geographic regions they traced their ancestry. 23andMe provided those results to their customers after the samples were processed on Illumina machines. Illumina soon formed partnerships with a growing list of these personal genomics companies, offering the technological underpinnings to generate the products that they sold to people curious about their own DNA.

Risch and Merikangas recognized in 1996 that the only thing standing in the way of the widespread use of genome-wide association studies were the technological barriers associated with the cost and time involved with processing the enormous data. Innovations like the bead array removed those barriers, and a flood of studies followed. Only a couple were published by 2005, a handful more in 2006. But in 2007 the number jumped to nearly 100, and then over 150 more in 2008. After the linkage studies failed to reliably find genes responsible for common diseases, the geneticists gravitated toward the proposal to instead embrace the genome-wide association study methodology. There was almost a full decade between when Risch and Merikangas first suggested the shift and when the technology was available to deploy it in a general fashion. Would genome-wide association studies work? Would they turn up the genetic associations that the linkage studies failed to deliver? Sure enough, the growing pool produced thousands of hits, identifying genomic regions newly associated with Parkinson's disease, bladder cancer, anorexia, obsessive-compulsive disorder, addiction, and asthma. Indeed, it soon became difficult to find a human trait that hadn't been subjected to a genome-wide association study.

Risch and Merikangas's proposal worked, which was exciting news for the geneticists. The results of the genome-wide association studies, on the other hand, were disappointing, not at all what the geneticists expected. The medical genetics community hoped the methodology would pick up the handful of regions in the human genome responsible for common, complex traits like asthma, diabetes, and Parkinson's disease. The genes implicated, it was hoped, would then provide the foundation for prediction and medical intervention. Somebody with the risky versions might take preventive actions to avoid getting sick that somebody with the safer versions need not, or somebody with the risky versions might be prescribed a pharmacogenomic drug specially designed to target one or more of those risky genes—personalized medicine in action. The genome-wide association studies, however, did not deliver on that expectation. Instead of finding a handful of regions in the genome, each of which made a sizable contribution to the diseases under investigation, the methodology found dozens or even hundreds of regions in the genome, each of which made a minuscule contribution. The genetic effects for the vast majority of the studies were so small and distributed so diffusely that translating that scientific information into personalized medicine—actually making predictions and interventions for patients—became severely compromised.

Medical geneticists were genuinely flustered by what came to be called their "missing heritability problem," but they were not dissuaded. Sizable genetic contributions to common, complex diseases proved elusive, but headway was being made in domains like rare diseases and some cancers. What's more, they didn't think it necessary to give up yet on the common traits. The solution, they countered, was simply more data—much, much more. Either the common genes that they were seeking had even smaller effect sizes than they initially anticipated, or the genes that they were seeking had fairly large effect sizes, but they were extremely rare. Those two routes went by two different names: the "common disease common variant" hypothesis and the "common disease rare variant" hypothesis.

The idea of the former was that a common disease like Parkinson's was the product of many, many genes, each of which made up only a tiny but collectively substantial portion of the risk associated with developing the trait, and so the difference between somebody who was at particularly high risk of Parkinson's and somebody who was at particularly low risk was which combination of those common variants they carried around in their genomes. The idea of the latter hypothesis instead was that a common disease like Parkinson's was the product of genetic variants that actually had a substantial impact on an individual's risk but that were each exceedingly rare in populations. So now the difference between somebody who was at particularly high risk of Parkinson's disease and somebody who was at particularly low risk was whether they had one of the very risky and very rare mutations; moreover, ten people at high risk of Parkinson's might have ten completely different very risky and very rare mutations. Either hypothesis, the thinking went, salvaged the promises of personalized medicine when it came to the common, complex diseases. Sure, genetic prediction might not be as simple as tracking two genes, but the hundreds of genomic regions identified with the genome-wide association studies could all be mashed together to create a composite "polygenic risk score" for Parkinson's. Genetic intervention might not be as simple as giving everyone with Parkinson's disease the same pharmacogenomic drug, but perhaps different patients with different rare variants associated with Parkinson's could each get their own pharmacogenomic drug.

Either way, the route ahead spelled business for Illumina. Whether a geneticist wanted to chase down common genes with very small effects or risky genes with very small frequencies, the need was the same: bigger and bigger data sets. As the costs and time of genotyping continued to drop, genome-wide association studies with 1,000 or 10,000 participants quickly became obsolete, replaced by ones with 100,000 or 500,000 participants. There was even talk of 1 million participants. Moreover, while the genome-wide association studies could rely on sampling SNPs spread out across the human genome, finding the rare genetic variants worked best when entire genomes were sequenced. Want to scoop up information on 100,000 SNPs from 500,000 people? Illumina had an innovative product for that. Want to scoop up information on entire genomes from 10,000 people? Illumina had an innovative product for that too.

. . .

There is a sense in which the term "genomics" is quite young, and a sense in which it is rather old. It is young compared with "genome." That word was introduced in the 1920s and then received increasing attention after James Watson and Francis Crick discovered the double-helical structure of DNA in 1953. A genome is the whole set of nuclear DNA contained in an organism's chromosomes. Genomics, the study of genomes, came along in 1986. Scientists gathered in Bethesda, Maryland, that year to discuss mapping the entire human genome, a precursor to what would become the Human Genome Project. After the meeting, the group gathered at McDonald's Raw Bar, just down the street from the NIH campus, to discuss publishing a new journal devoted to their research on genomes. Most towns in Maryland have some version of the place—an unassuming neighborhood joint that serves superb lobster tail, crabs, and oysters. The scientists, over pitchers of beer, bounced around different ideas for the journal's name until Tom Roderick, a geneticist with the Jackson Laboratory, threw out "Genomics." Genetics was the study of genes, so it wasn't a stretch to see genomics as the study of genomes. *Genomics* appeared the following year, introducing a new scope of biological inquiry, the idea that something in its entirety—the whole genome, the whole set of nuclear DNA—could be its own object of study. As Roderick recalled his insight, "It was an activity, a new way of thinking about biology."

"Genomics" is old by another measure. There are now an untold number of self-described "-omics" disciplines—microbiomics, proteomics, metabolomics, foodomics. These started appearing and then proliferating about ten to fifteen years after the term "genomics" appeared. As the Human Genome Project was coming to its conclusion and receiving increasing public attention, other life science practitioners wanted to get in on the big-data revolution. If genomics could draw on technologies developed by companies like Illumina to study entire genomes, these other disciplines hoped to do the same thing with technologies that allowed them to probe their own objects of study in their entirety. Immunologists took on antibodyomics, the study of an organism's whole set of antibodies; neuroscientists embarked on connectomics, an effort to map all the neural connections of the brain; linguistic and cultural phenomena could be quantitatively tracked across centuries with culturomics. Adding "-omics" to the end of a discipline proudly signaled that it could do big-data science too, that it was also incorporating technological innovations to more quickly and more cost effectively accumulate more data.

Few life sciences avoided this omification in some form or another. Even environmental health scientists eventually found their colleagues talking about the "exposome," characterized as the whole set of environmental exposures that a human encountered over the course of her life from zygote to death. Exposomics attempted to mold the environment into something that could be studied just as the genomicists were studying genomes. Genomics had its SNP detection and sequencing technologies; investigating the exposome would require its own exposure biology tech. Genomicists had their genome-wide association studies; some environmental health scientists started conducting what they called "environment-wide association studies." Genomics was able to get all the data it needed out of drops of blood; exposomics would similarly find ways to measure somebody's history of environmental exposures based on biomarkers in their blood.

The NIH's Genes, Environment, and Health Initiative had two sides to it. Collins used his portion to contribute to the pool of genome-wide association studies ballooning in the medical genetics community. David Schwartz, head of the National Institute of Environmental Health Sciences, used his allocation to get the NIH in the exposome business with the Exposure Biology Program. There, they funded efforts to develop new technologies that could be deployed to measure environmental exposures on a big-data level, efforts like the artificial nose. Schwartz and Walt even had conversations about what the bead array could bring to the exposure biology community. It turned out, though, that by the time the Genes, Environment, and Health Initiative got up and running, Walt was already migrating away from the artificial nose work. Illumina, for its part, had abandoned it years earlier.

The Walt team, with graduate students like Keith Albert and Shannon Stitzel replacing Dickinson after he left for Illumina, initially took the optical nose into exciting new directions inside the laboratory. They worked to train bead array sensors to detect nitroaromatic compound vapors in the laboratory, which could indicate that explosives such as land mines were in the vicinity. They also used it to distinguish French roast coffee from Colombian or hazelnut grounds. Stitzel helped create a massive model of a dog's nose, from which the team learned how airflow and the distribution of sensors across a surface area could improve odor discrimination. After 9/11, the defense industry scrambled to find better ways to detect explosives. Consumer goods like coffee and alcohol could be monitored for quality assurance purposes. Environmental monitoring was possible on a big-data scale. Other labs claimed their artificial noses could detect lung cancer just by sniffing a patient's breath. The future of the optical nose specifically and artificial noses generally seemed bright.

But it's one thing to get the optical nose to work in the lab, and it's another thing to get it to work out in the world. Albert, supported by a team of engineers and scientists from Lawrence Livermore National Laboratory and the University of South Carolina, strapped one device, what they called their "portable sniffer," to something that resembled a baby stroller and took it to Fort Leonard Wood. The military complex, created in the 1940s as a training location in anticipation of the United States entering World War II, is spread out over about a hundred square miles of Missouri's Ozark Mountains. After the Vietnam War, the base shifted toward engineering operations, but it still has all the features of a combat training facility: shooting range, obstacle course, fields for explosives instruction. It's a brutal climate; winters are frigid and wet, while summers are oppressively hot, muggy, and buggy.

The task for Albert was to see if the Walt team's portable sniffer could reliably detect land mines. The base staff dispersed land mines with the triggers removed on two plots of land, one a trial field where the locations were known to the researchers and one test field where the locations were hidden. After the optical nose was trained on the known locations, the plan was to see if it could detect the hidden ones

The "portable sniffer" at Fort Leonard Wood, Missouri.
Photograph by Keith Albert

in the test field. The removal of the land-mine triggers ensured that Albert and the group's contraption wouldn't set off the explosives, though they were still warned to evacuate the area in a hurry if a lightning storm approached. It was grueling work even when there was no lightning. Albert arrived in the summer, but he couldn't wear sunscreen for fear that the vapors from it would throw off the optical nose. He also couldn't use bug spray for the same reason. In the end, the portable sniffer never made it to the base's test field because they couldn't even get it to detect land mines in the trial field where the locations were known. The heat and the humidity and the dust, combined with the fact that the land mines were buried under soil, outmatched the bead array.

The bead arrays built into Illumina's BeadChips and BeadStations had a distinct advantage over the bead array built into Walt's portable sniffer. For the genetic applications, keep in mind, the biological samples, in effect, traveled to the devices. The machines could be maintained in temperature- and humidity-controlled spaces with special air filtration systems that kept outside contaminants to a minimum. The samples, moreover, were prepared and introduced in a controlled environment. For environmental applications, on the other hand, the devices needed to go to the samples; the whole system must work outside among contaminants. You can't control the environment when the purpose of the device is to detect substances in the actual, uncontrolled environment.

Portability was just one of the challenges that the environmentally purposed bead array faced compared with its genetic counterpart. A genome is molecularly immense, but as an object of chemical scrutiny, it's pretty straightforward. There is a long sequence of nucleic acid base pairs, and at each spot there are four possibilities. The scale is big, but the chemical complexity and, in turn, the sensor detection challenges are constrained by the limited chemical variables. The environmental applications involved vastly more complex analyses. A sensor device must be attuned to many different possibilities. Each additional item added for detection made the technological challenges more difficult. Furthermore, with environmental exposure data, a researcher often wants to know the degree of an exposure, not just whether an encounter occurred at all; there are safe and unsafe levels of certain chemical compounds when it comes to human health, so it's not enough for a sensor to conclude "present" or "absent." It's far easier for a sensor to detect presence/absence rather than degree, which is why Walt's optical nose gravitated toward applications where presence/absence sufficed, as with land mines, where there's no safe level. An environmental sensor must also be able to detect very small quantities of the thing to be analyzed, amounts that are common in nature but far more difficult to detect and measure. Finally, the environment is constantly changing, and an environmental health scientist wants to know how those fluctuations affect health. That means data needs to be repeatedly collected, and repeated measurements are much harder on a manufactured device.

Those were the practical challenges. There were financial ones, too. Walt contacted multiple large companies to feel out their interest in investing in the optical nose. The Dow Chemical Company could use the device to ensure that its tanker trucks were devoid of some volatile compound before adding a new product. Cargill, the global food and beverage corporation, could employ the optical nose to make sure that the oranges it was receiving from its suppliers were of the right flavor and did not have bitter components that would impart a bad taste. The companies, however, were not interested in paying for the research and development of the technology. In contrast to the pharmaceutical and personal genomics companies as well as the academic and federal biomedical researchers who were racing to get more and more genetic data and so willing to invest heavily in efforts that drove down the costs and time, the market wasn't there for investing in the optical nose. Walt, facing this practical and financial reality, eventually turned his sights and the efforts of his graduate students elsewhere.

Illumina saw the same dynamic as Walt but on an accelerated, private industry schedule. The Illumina scientists eventually decided that the optical nose tech was years away from being ready for commercialization, whereas the genetic applications of the bead array were deployable much sooner. Likewise, they realized they would have to go banging on doors to get potential customers interested in what they created with the optical nose, whereas the customers for the genetic applications were coming to them. The group ultimately dropped the optical nose program altogether. By the time Illumina went public in 2000, it was unmistakably a genomics technology company.

The vision for the NIH's Genes, Environment, and Health Initiative—the guarantee to politicians like Secretary Leavitt—was that the Genome Institute's genome-wide association studies and the Environmental Institute's exposure biology innovations would come together and reveal the great mysteries regarding how genes and the environment interacted to produce human health and illness. The problem was that the technologies available to the Genome Institute and the technologies available to the Environmental Institute were worlds apart. That tension was on full display at the National Press Club event in February 2006. It was easy for Collins to tell the reporters about doing genome-wide association studies because there were already several SNP detection technologies on the market. Schwartz, on the other hand, had to ask the reporters to imagine how wonderful it would be if the sensor hanging off his lapel could detect a dozen different exposures. The exposome research was just trying to get up and running. There were some signs of progress in the wearable sensor domain. Fitbit, for example, went live in 2007 with its activity trackers at the same time that 23andMe was advertising its genetic testing. The smartwatches that soon followed can do some big-data things very well—count your steps, monitor your heart rate—because those things are fairly easy to measure and quantify, but they can't tell you if you'd crossed paths with benzene or inhaled sulfur dioxide or drank water contaminated with fuel. That's what Schwartz's Exposure Biology Program was supposed to provide. The equivalent would have been if the Genome Institute were tasked with discovering the bead array technology themselves and then also figuring out some way to mass-produce the BeadChips, BeadLabs, and BeadStations without a company like Illumina to finance and develop it all. That, of course, would have been a ridiculous expectation, and yet the thought that the Environmental Institute's Exposure Biology Program could keep up with the Genome Institute's genome-wide association studies created an equally impossible expectation on the environmental scientists. The result was two separate efforts paired under the same funding umbrella. The idea that the genetic results and the environmental results would come together and elucidate gene-environment interactions was a mirage.

. . .

More than five thousand babies in the United States died from sudden infant death syndrome (SIDS) in 1991. Five thousand parents or caregivers had laid to sleep a happy and healthy infant, only to horrifically find a lifeless body hours later. SIDS was the leading cause of infant mortality in the United States that year, accounting for more than one-third of all deaths among babies who did not reach their first birthday. It was a terrifying public health threat.

History and literature are replete with tragic accounts of "crib deaths." SIDS, as a diagnostic category, was introduced in 1969. It is not so much a condition as the lack of a condition. A SIDS death is one that remains unexplained even after an autopsy, an examination of the death scene, and a review of clinical history rule out all other potential causes, such as a cardiac event or an injury. Much was known about the trends in these crib deaths by the mid- to late twentieth century: they were more common during the winter; the riskiest developmental period was between two and four months; the deaths typically occurred in the middle of the night. But what actually caused the heartbreaking losses of life remained elusive. Theories abounded as to the causes: they were the result of an enlarged thymus gland, or an allergic reaction to cow milk, or a vaccine, or a mother's cocaine use during pregnancy, or viral infections.

The New York pediatrician Harold Abramson in 1944 had pointed to a far simpler explanation—a baby's sleeping environment. Abramson noted a worrying rise in "accidental mechanical suffocations" over the previous decade. Reviewing 139 infant deaths between 1939 and 1943, Abramson pointed to a series of features surrounding how a baby was put to bed as the culprit. Loose bedding, a crib cluttered with toys and pillows, and placing a baby on its stomach all were found most often among the cases of crib deaths. The solution, Abramson advocated, was a large-scale public outreach campaign aimed at educating parents about a baby's safe sleeping environment—just the baby in the crib, kept warm but not hot, sleeping on her back.

Abramson's recommendations, alas, were largely dismissed by the child health community for more than four decades. Blaming the sleep environment was tantamount to blaming the parents, critics charged, an accusation far too cruel for parents already in the depths of grief. There were obvious dangers associated with babies sleeping on their backs, it seemed; they spit up often and so could choke easily. Surely there must be a straightforward biomedical cause yet to be discovered. But not until the 1980s did pediatricians come around to the wisdom of Abramson's words. Clues came from international studies that revealed different rates of SIDS associated with different cultures of putting babies to sleep. Parents in China, for example, very rarely encountered the horror of a baby who'd succumbed to SIDS; the British pediatrician David Davies connected that to the fact that Chinese parents typically put their babies to sleep on their backs. Babies in the Netherlands traditionally slept on their backs too. Alas, at a large child health conference in 1971 several scientists advocated for the virtues of tummy sleeping. Dutch researchers went back and scrutinized the rates of SIDS over the previous decades and found a sharp increase within their population after that 1971 event. Other retrospective reviews of data regarding SIDS also turned up prone sleeping as a common factor among the infant deaths.

Those observational studies were enough to spark a series of hypothesis-driven, prospective cohort as well as case-control studies around the world designed to explicitly probe the impact of sleeping position on SIDS. Researchers began following babies from birth, collecting information from parents about all sorts of details concerning the babies' environments—how they were positioned at night, what was in the crib, whether the parents drank or smoked, what the babies were fed, the socioeconomic status of the family. The evidence that emerged was clear; babies laid prone were far more likely to die of SIDS than babies laid supine.

Once the pediatric community came around to Abramson's insight, it also heeded his advice regarding public health education. In the early 1990s, Duane Alexander and his National Institute of Child Health and Human Development took the lead in the United States alongside the American Academy of Pediatrics in launching the Back to Sleep campaign, a nationwide outreach effort aimed at providing new parents with advice on how to lay their infant to rest. It was in line with Abramson's advice from decades earlier. The Child Health Institute also launched its National Infant Sleep Position study, telephoning parents across the country to see how well the public health message was being received and implemented. Alexander, for his part, took to the front lines. He spoke with reporters. He appeared on television interviews. When a caller phoned in to Alexander's appearance on C-SPAN, she sternly told him, "You know, I don't believe these statistics at all," pointing to her experience with five children who all slept happily on their stomachs, as well as fears of choking that could follow back sleeping. Alexander, like a pediatrician hearing out an anxious parent's vaccine fears, listened patiently, validated the caller's worry about suffocation, and then calmly explained the data. Before they embarked on the Back to Sleep campaign, he assured her, they looked to see if other countries that switched to back sleeping saw an increase in chokings and infant suffocation. They did not. Supine sleeping was not dangerous, and it saved lives.

The results of the campaign and the associated research were truly impressive; in 1996, just a few years after the Back to Sleep campaign launched, the percentage of parents who placed their babies to sleep on their stomachs fell from 70 to 24 percent. SIDS deaths, in turn, plummeted too. At the same time, the pediatric researchers learned both that SIDS deaths remained stubbornly high among Black American and Native American babies and that Black American and Native American parents were switching from the prone to the supine sleeping position more slowly; as a result, targeted education efforts were developed specifically to reach those communities.

The actual cause or causes of SIDS remain obscure. The leading contender at present is the triple-risk model, which hypothesizes that a baby with some underlying biological risk (like a brain dysfunction associated with respiratory control) reaches a particularly risky period in development (like the two-to-four-month period) and then is exposed to a risky environment (like being laid to rest on her tummy). What is not under debate is the positive health impact of putting infants to bed on their backs. Just one decade after the Back to Sleep campaign rolled out, SIDS deaths in the United States dropped by close to 60 percent. It didn't eliminate the tragedies, but it greatly reduced them.

Also indisputable is the fact that the scientific research that paved the way for that public health triumph was decidedly low-tech. The most advanced device utilized was a telephone. The data collected did not require technically manufacturing synthetic oligos; it required accumulating detailed information about crib, home, and family environments. The sample sizes were not in the hundreds of thousands; they were in the hundreds and sometimes thousands. The effort was not about replacing one-size-fits-all health care with personalized drugs from pharmaceutical companies; it was about identifying features of children's environments that genuinely were one-size-threatens-all and could be reduced with something as simple and inexpensive as laying a baby to rest on her back.

The epidemiological research that led up to the Back to Sleep campaign, the insights that Philip Landrigan derived from his studies of lead exposure among the children living in Smeltertown, the lessons of Janet Hardy's work with the Baltimore mothers and their babies as part of the Collaborative Perinatal Project—these were the scientific inspirations for the architects of the National Children's Study, not the desire to conquer the exposome. Alexander wasn't opposed to incorporating new technological innovations into the project. In fact, the study included a category of efforts called "formative research projects," which were specifically intended for testing out new and innovative ways of measuring the environment. But there wasn't a sense that a properly conducted nationwide study of children's health needed to sit idle until technologies like Walt's artificial nose or Schwartz's lapel sensor were ready for prime time. Traditional epidemiological research was capable of identifying threats lurking in the environment as long as the right data was collected in the right way. That often took some time; it tended to cost money; it generally didn't promise profits for private industry. But it got results: it saved lives.

Once Congress provided the funds to support the National Children's Study in 2007, it was time to invite an external review of what was being planned. Child Health asked the nonprofit National Academies of Sciences, Engineering, and Medicine, the same organization that sponsored Landrigan's 1993 report *Pesticides in the Diets of Infants and Children,* to convene an independent panel of experts to evaluate the design of the study—the sampling of children, the plan for recruiting the families, the hypotheses that would be tested, the data that was to be collected, the results that were to be returned to the participating children and their parents. It was a significant undertaking. The dozen experts in public health, sociology, psychology, human development, environmental health, education, medical ethics, and statistics met several times over the course of a year, processing the detailed research plan and discussing its potential and limitations.

The report that the group released in the spring of 2008 called the children's health initiative a "landmark study" that presented science with an "unparalleled opportunity to examine the effects of environmental influences on child health and development." The reviewers praised the extent to which the study leadership responded to the mandate set out in the Children's Health Act of 2000. The basic design of the study—100,000 children selected with a national probability sample and followed for more than two decades—was highlighted as a major virtue, permitting the careful scrutiny of biological and environmental contributions to children's health as well as the social and psychological influences on health generally and health inequities specifically. The panel made several dozen recommendations, many of them encouraging the National Children's Study to be even more ambitious—more frequent in-person meetings with the families, more data about things like the occupational exposures of the fathers and the legal immigration status of the parents, more attention to the impacts of access or lack of access to medical care and child care, a more substantial pilot phase to test out the recruitment process and field deployment of data collection.

The response from Child Health was a mix of gratitude and a sense of validation. In some cases, the recommendations made by the panel aligned with developments already under way in the National Children's Study; the study had continued to grow during the year that the National Academies panel convened, and so it turned out that the study leadership had independently hit on some of the very same ideas as the panel. The pilot testing to be undertaken by the Vanguard Centers, for example, had already been extended by six months in order to provide them with more time to try out the door-to-door recruitment and the various tests to be administered in the homes. Other recommendations were found problematic. The more frequent home visits, for instance, would have excessively driven up costs; likewise, the Vanguard Center directors warned against trying to collect information on the immigration status of parents because it could scare away potential participants who were legally vulnerable.

While the National Academies report was circulating among the study community, the Vanguard Center directors were preparing to kick off in their communities. The plan was for the centers covering Queens in New York City and Duplin County, North Carolina, to begin in January 2009, followed by the other five sites in April that same year. In the lead-up to the actual recruiting, the principal investigators of the Vanguard Centers were fostering the community connections that had to be in place if recruitment was to be successful. Landrigan, in New York, took stock of the massive apartment buildings his recruiters would have to cover. In Utah, Edward Clark was meeting regularly with the Navajo Nation's Tribal Council. The council needed to endorse the study if members of their community were to participate, and that meant explaining the value of the study—for children generally and Navajo youth specifically—and responding to any concerns that the community raised.

The extended pilot testing would delay things, and that created challenges. The Vanguard Centers were tasked with judging the practical feasibility of everything that had been proposed by the National Children's Study working groups so as to home in on the set that was most scientifically valuable and affordable. Until then, precisely budgeting the costs of the study years down the road was largely an exercise in guesstimates because it remained to be seen what they'd learn from the pilot testing. Peter Scheidt and Sarah Keim, in the program office, were confident about what the next year would cost, but the study was expected to run for two decades, so assigning annual costs up to twenty years out was simplified to budgeting them all at $99.4 million a year. When a budget officer in the Child Health Institute picked up on this, Keim replied, "Unfortunately, we are all sticking to some oversimplified assumptions about the cost of the Study (like that it will cost $99.4m per year for the last approx 20 years of the study)." That was an uncertainty which would need to be tolerated until the Vanguard Centers refined the final study protocol. As the National Academies review emphasized, much of the study's success or failure hinged on getting it right from the beginning, and so it was important to take the time needed to make that happen.

The National Children's Study community was extremely optimistic about the project in 2008, at least from a scientific and logistical perspective. The Vanguards were gearing up. With the funds allocated by Congress, the hosts of the next batch of "Wave 1" sites were identified in places like Wayne County, Michigan, and Providence County, Rhode Island. The National Academies gave the research plan an independent stamp of approval. The great impediment to the National Children's Study, as federal scientists like Alexander and Scheidt as well as academic scientists like Landrigan and Clark saw it, was political. They had no serious advocates within the Bush administration. And the conservative president's choice of NIH director, Elias Zerhouni, showed no signs of support for the study once the collaboration with Collins deteriorated years earlier. That changed in November 2008, when a Democrat and Black American was elected to be the next president of the United States. Barack Obama, in the child health experts' view, represented a political sea change—a more sympathetic administration and a new, presumably more environmentally friendly director of the NIH on the way. With that final, political piece of the puzzle falling into place, Alexander decided to make the case for a far grander National Children's Study.

5

The Politics of the Personal

ON NOVEMBER 14, 2008, just days after Barack Obama was elected president, the outgoing secretary of health and human services, Michael Leavitt, left what he called a "note on the desk" for the next administration. Leavitt's note, at 302 pages long, would have taken some time to read. In it, he encouraged the Democrats replacing Republicans to continue investing in the promises of personalized medicine. Leavitt, who started making personalized medicine a priority for the Department of Health and Human Services in 2006, called the two years that followed a "prologue" for personalized medicine in America. Now it was time for the Obama administration to write the next chapter.

The president-elect was a particularly receptive audience for Leavitt's message. In his short tenure as a U.S. senator, Obama was a leading advocate for genomic medicine and the importance of federal leadership on the new approach to health care. That Leavitt, a Republican, and Obama, a Democrat, would see eye to eye on the value of personalized medicine was no anomaly. In the hyper-partisan political environment that firmly gripped the United States by 2008, federal investment in medical genetics research was among the few endeavors that conservatives and progressives alike openly embraced, albeit for different reasons.

This was in stark contrast to the political landscape surrounding federal investment in environmental research. Richard Nixon created the U.S. Environmental Protection Agency in 1970 and called the 1970s the "decade of the environment," launched by the nationwide Earth Day celebrations that took place that spring. Widespread political support for federal legislation like the Clean Air Act and the Clean Water Act were championed by both Republicans and Democrats. Just a few decades later, however, the dynamic had radically changed. Republicans routinely assailed "environmental extremism" as an anti-industry, anti-liberty, anti-American ideology and employed various methods—sometimes subtle, sometimes brazen—to stifle federal support for the scientific research that advanced it. George W. Bush handed Obama an EPA that was financially strapped and institutionally demoralized. Environmental regulation and protection, and the environmental research that undergirded those efforts, had become among the more polarizing issues in American politics.

Transitions from one presidential administration to the next, especially when they involve a switch in political party, are moments of tremendous upheaval across the federal government as new agendas replace old and new personnel are selected to implement a vision for the country. The National Institutes of Health is one of the many federal agencies impacted by that process, most directly when the new president nominates a new director. Duane Alexander, head of the NIH's Child Health Institute, and Francis Collins, director of its Genome Institute, both maneuvered in 2008 to take advantage of the shifting political landscape, with profoundly different levels of success.

. . .

How did the Republican Party go, in just a few decades, from one that created the EPA to one that intentionally and systematically tried to hobble it? To understand that evolution, it's important to recognize that while Nixon formed the new federal agency in 1970, he was never a committed supporter of environmental causes. The famously calculating president, keeping his eye on the presidential election of 1972, noted a passion for the environment that only grew across the nation after *Silent Spring* was published. Nixon feared a Democratic challenger would try to ride the grassroots wave into the general election, and he wanted to cut off the threat. Prior to 1970, most air and water pollution concerns were handled by the states, and the federal research and regulatory infrastructure that was in place pertaining to the environment was spread out across many different sectors: the Department of the Interior, the Department of Agriculture, the Atomic Energy Commission. Nixon's EPA wasn't a massive federal investment in the environment so much as it was a massive federal reorganization of existing systems pertaining to the environment. Still, the restructuring did empower the new agency with a unified mandate, and when that centralization of authority was combined with federal legislation from Congress, it put the pieces in place for an influential new face of federal environmental research and regulation. William Ruckelshaus, the first administrator of the EPA, took to the job, helping implement the Clean Air Act, bringing to life the agency's enforcement arm tasked with seeking out and legally addressing violations, and advocating for the ban on DDT.

After Nixon was reelected, he became far less enamored of environmental concerns. He was convinced there was nothing he could do to satisfy environmentalists; moreover, the actions taken to appease them simultaneously angered the energy, manufacturing, and agricultural industries that were subjected to the new scrutiny and regulations. Nixon, for example, was furious about Ruckelshaus's efforts to eliminate DDT, fearing an insurrection from his bloc of farming constituents. The ban proceeded, but Ruckelshaus was soon shuffled to the FBI, after which the EPA lost momentum. Nixon's administration set about financially constraining the agency and limiting the reach of the legislation with delays and exemptions. That persisted until Nixon left office.

The 1970s brought a stagnating economy and rising inflation along with an emerging cohort of influential conservative think tanks, such as the Heritage Foundation and the Charles Koch Foundation. These organizations united around the idea that America's economic problems were caused by government overreach that stifled personal initiative and success. The federal government, they charged, was less efficient than private industry; it hampered free markets and job growth with unnecessary regulations; and it inappropriately intruded on the individual liberties of its citizens. Federal efforts to protect the environment and monitor the health impacts of an unsafe world were by no means the only target of this burgeoning neoliberal philosophy, but they were a central focus. The grassroots environmental movement that took shape in the 1960s was reconceived as an ideological overreach driven by intellectual elites and harmful to good business and working-class Americans.

No national figure capitalized on this growing sentiment more effectively than Ronald Reagan, who was elected president in 1980 on a platform organized around the idea that the federal government was the problem and reeling in its regulatory activities was the solution. "The nine most terrifying words in the English language," Reagan liked to say, were "'I'm from the government, and I'm here to help.'" Reagan's administration used executive actions to render the federal government's regulatory powers impotent. He appointed as EPA administrator Anne Gorsuch, a former lawyer and state legislator with no significant managerial experience or environmental expertise but someone who did pledge her loyalty to Reagan's governing philosophy. Within just a year, Gorsuch welcomed drastic budget cuts to the agency, particularly hamstringing its research and development efforts, thereby limiting its ability to detect new environmental threats or update the science on existing ones. She also significantly downsized its workforce, firing or demoting staff and leaving others so demoralized that they quit.

The Reagan administration's transparent assault on the EPA, reported in the media, led to a congressional investigation. Gorsuch was the first cabinet-level official in the nation's history to be found in contempt of Congress when she didn't cooperate with their inquiry. The investigation eventually revealed widespread neglect, mismanagement, and corrosive conflicts of interest at the agency. Throughout 1983, Gorsuch and dozens of other EPA officials either resigned or were fired; one even went to prison for perjury. The humiliating rebuke left the administration on its heels. Reagan, in desperation, brought Ruckelshaus back to administer the EPA, hoping to minimize bad press before the 1984 election. In his second term, Reagan took a far more conciliatory approach toward dealing with the environment. His successor, George H. W. Bush, tried to push back against the Reagan pattern, signing the United Nations Framework Convention on Climate Change and the 1990 Clean Air Act amendments to address acid rain, but Bush was fighting an uphill battle against most of his fellow Republicans.

By the time Bill Clinton left office in 2001, the political divide surrounding the environment was deeply entrenched. Whereas emboldened Republicans, inspired by Newt Gingrich's "Republican Revolution" in 1994, returned to a more antagonistic approach to environmental research and regulation, the Clinton administration took a number of steps to affirm the Democratic Party's commitment to it: an executive order inspired by the environmental justice movement that demanded the EPA and other federal agencies identify and address the disproportionate impact of environmental policies and regulations on marginalized communities, a severe reduction in the amount of arsenic permitted in drinking water, a set of new restrictions on diesel engine emissions, and an increase on the scrutiny of lead paint in old buildings, to name just a few initiatives.

In 2000, during the presidential campaign, George W. Bush virtually ignored Al Gore's record of environmental activism. Gore had built his political career in part on championing environmental causes, in 1992 publishing his book *Earth in the Balance,* a call to arms regarding climate change, and then taking a coordinating lead on environmental initiatives in the Clinton administration as vice president. When Bush took the White House, he shifted the federal government back toward less environmental regulation, protection, and research. The Bush administration had learned enough from the Reagan debacle not to assault the federal environmental research and protection mechanisms overtly. Instead, it took a subtler approach, quietly rolling back Clinton-era environmental regulations, packaging industry-friendly priorities in seemingly pro-environmental language like the "Clear Skies Initiative," reintroducing heavy budget cuts to the EPA, making the scientific review process in the agency less transparent. Nevertheless, a clear dynamic had become plain for all to see by the end of Bush's presidency: Democratic presidents tended to embrace environmental causes and prioritize federal environmental research and regulation, while Republican presidents demonstrated an aversion toward committing the federal government to environmental management, opting instead for solutions that gave private industry more control and flexibility.

The National Children's Study, from its very origin, bore the marks of this partisan history. Carol Browner, head of the EPA during the Clinton administration, made a concerted effort to steer the agency toward focusing on populations and communities that had traditionally been ignored. The environmental racism spotlighted in the 1980s by the Black protesters of Warren County, North Carolina, laid the foundation for President Clinton's 1994 executive order that required all federal agencies to identify and address any actions that they might have been facilitating which took a disproportionate toll on minority and low-income populations. The 1993 National Academies report *Pesticides in the Diets of Infants and Children,* led by Philip Landrigan, contributed to an executive order from Clinton that called for the creation of a federal task force committed to investigating environmental health and safety risks affecting children, co-chaired by Browner and timed to coincide with the 1997 Earth Day celebrations. That task force had a number of workgroups organized around environmental impacts on various pediatric health conditions: asthma, cancers, unintentional injuries, developmental disorders. Alexander, by then one of the federal government's leading voices on developmental disorders, co-led the group focused on that topic, and it was during a 1998 gathering in his office, when the federal scientists were imagining big initiatives the task force could propose, that one of them reminded the group it had been forty years since the Collaborative Perinatal Project. Perhaps this was the moment to encourage the federal government to take the lead on launching a new, nationwide, longitudinal study of the country's children.

Alexander took the proposal back to the larger task force, which embraced the idea, making it one of the key recommendations to come out of the environmental health effort. That proposal, in turn, became a central piece of the Children's Health Act of 2000, among the last pieces of legislation that the Democratic president signed. Clinton hailed the longitudinal study at the heart of it as "a national long-term study of environmental influences on children's health and development, that will provide critical information about environmental, social, and economic factors that affect children's health."

When George W. Bush replaced Clinton in the White House, administrative support for the study evaporated, culminating, as we've seen, in the administration's call for terminating the project.

Congress, of course, threw a financial lifeline to the study, but Bush's budget proposed eliminating the study again in FY2008 and once more for FY2009, with the NIH director, Elias Zerhouni, arguing as late as September 2008 that there were far better uses of federal research dollars than the children's study. Democratic Congresses kept delivering financial support in spite of the Republicans.

Those developments over the previous decades provided the backdrop through which the National Children's Study leadership viewed Barack Obama's victory. Alexander took stock of the recent review from the National Academies alongside the fact that the Vanguard Centers were about to begin recruiting for the study, and believed that a president with more affinity for environmental concerns was about to take over the White House and appoint a new NIH director who would presumably be far friendlier to the environmentally focused study. Zerhouni stepped down as NIH director at the end of October, handing the reins to his deputy Raynard Kington, who was serving as acting director of the NIH until Obama nominated, and the Senate confirmed, a new leader.

Shortly after Obama was elected, Alexander went to the NIH's budget office and asked if they could request that the National Children's Study allocation be included in President Obama's budget for the coming year, rather than forcing Congress to once again add it after the fact. When that request was approved—a good sign of newfound political support—Alexander requested one more thing. The total projected cost of the National Children's Study had been artificially held low by the budget directive that existed under the previous administration, he explained, so he wanted to talk with Kington about the political feasibility of revising upward the $3 billion number. The request for that conversation, to Alexander's relief, was also approved, and the Child Health director was encouraged to come to the meeting with data and specifics. Alexander's optimism that, given enough time, solutions to even the most difficult problems often presented themselves appeared to be validated once more.

Peter Scheidt and Sarah Keim, in the program office at Child Health, put together a menu of proposals for the National Children's Study that Alexander could take to Kington: the pared-down version they'd long been reporting that would come in at $3 billion, mid-level versions that would cost between $4 billion and $5 billion, and a "Cadillac model" that included everything the hundreds of scientists associated with the National Children's Study had been envisioning with a price tag of more than $7 billion. "I hope the meeting goes well," Keim told Alexander as he left the Child Health Institute and headed toward Building 1, where the Office of the Director was located.

. . .

Many of the features of environmental research that elicited disdain from political conservatives were the very same features that attracted them to personalized medicine. Whereas environmental research came to represent a threat to private industry, personalized medicine instead grew directly out of companies like Genaissance, Pfizer, and Illumina. The plummeting costs of genetic sequencing and genotyping were taken to be clear examples of what can happen in biomedicine when corporate competition plays out in a free market.

Personalized medicine, aside from being business-friendly, was also welcomed by conservatives because of its straightforward embrace of individualism. Advocates argued that personalized medicine reoriented medicine around the person, which ideologically resonated with neoliberals. Rather than approaching all patients as if they were identical, personalized medicine purported to treat each patient as special. The unique genome of each person represented a biological embodiment of individual liberty, a biochemical distinctiveness that could counter the collectivist notion that humans are generally very much the same when it comes to what makes us sick or healthy.

Secretary Leavitt started making personalized medicine a priority for the federal government in 2006, with the rollout of efforts like the genome-wide association studies at the heart of the Genetic Association Information Network and the Genes, Environment, and Health Initiative. A report released by Health and Human Services just a year later counted more than fifty programs spread out across the federal government designed to facilitate the development of personalized medicine. The new approach to medical care, according to Leavitt, empowered patients, putting them in control of their own health and health care. Practically speaking, the patient's role in medicine changed very little with the addition of genomic information; a physician still decided whether some diagnostic test was appropriate and still ordered the prescription drug deemed a good match. But because it was based on the patient's own DNA, personalized medicine gave the illusion that it shifted power toward the patient. It gave them, the argument went, ownership of and also responsibility over their health; health care was said to shift from something that happened to patients to something that patients took charge of guiding with their own genetic information. The role of the federal government, for Leavitt, was creating a landscape that unleashed the power of cost-effective, efficient, free-market competition that would allow personalized medicine to realize its potential, and so he prioritized streamlining the drug approval regulatory system, overhauling the health-care reimbursement process, protecting intellectual property associated with biomedical discoveries, and facilitating ongoing collaborations between public science and private companies that thereby integrated federal activities with industry interests.

It wasn't just Leavitt. The President's Council of Advisors on Science and Technology, an advisory body to the president, also produced a report encouraging the prioritization of personalized medicine for President Bush in 2008. And the Republican Party platform in 2008 even incorporated language right out of the pharmacogenomics marketing playbook, promising to swap out the "cookie-cutter" treatment of patients with health care that "will personalize and coordinate their care to ensure they receive the right treatment with the right health care provider at the right time."

President Obama didn't need to be convinced about the value of personalized medicine, since he had sponsored the first legislation in Congress devoted to it. According to Obama's health policy adviser, Dora Hughes, he was naturally inquisitive and interested in science and technology; plus, losing his mother to cancer when she was only fifty-two gave him a personal investment in the topic. Hughes sensed the growing excitement surrounding genomic medicine in the middle of the first decade of the twenty-first century, and so she approached the senator about sponsoring a bill that would promote personalized medicine, to which he was receptive.

Senator Obama introduced the Genomics and Personalized Medicine Act in 2006. Its provisions bore some resemblance to Leavitt's Republican priorities; for example, Obama's proposal included tax incentives aimed at private companies developing diagnostics alongside their pharmaceutical compounds, and also called for updating the U.S. Food and Drug Administration's process of approving those diagnostics. But it also took a decidedly progressive tone. The purpose of the bill wasn't just to create a business-friendly environment for personalized medicine. It was about access, "to secure the promise of personalized medicine for all Americans." "Genomics holds the promise of revolutionary advances in medicine," said Obama upon announcing the legislation. The question was, how could the federal government help give everyone safe and effective access to that revolution? In his legislation Obama wanted to incorporate a more diverse population of research participants, to prioritize attention to health conditions impacting minority populations, to standardize the review and approval of genetic tests, and to protect consumers from shady companies that used questionable genetic tests designed to sell pseudoscientific gimmicks.

Obama said little about personalized medicine on the presidential campaign trail in 2008, where it was crowded out by more pressing matters like the U.S. military occupation of Iraq and the Great Recession. Still, advocates of personalized medicine knew that they had a political ally in Obama. They sensed a real opportunity in his election—a "personalized medicine president" in the White House. Few knew this better than the director of the NIH's Genome Institute.

Collins navigated his time in Washington, D.C., with extraordinary political dexterity. His open embrace of religion appealed to political conservatives, and his expanding reputation as a face of American biomedical research endeared him with science-minded progressives. Politicians from both parties asked the charismatic scientist if he brought his guitar along when he visited Congress to update them on the Genome Institute's budget, which lightened the mood when he was asking them year after year to keep funding the Human Genome Project. What's more, Collins pressed hard to persuade Congress to pass legislation that prevented genetic discrimination, which could have become a matter of public concern, deriving from fears that somebody might be denied health insurance or fired from a job based on the results of a genetic test. There was little evidence in the 1990s and first decade of the twenty-first century that genetic discrimination was actually a problem, but advocates of personalized medicine like Collins worried that public anxieties about just the potential for such disparate treatment could be enough to dissuade Americans from participating in the science and medicine, and so they encouraged elected officials to alleviate those concerns by proactively prohibiting such discrimination. Congress ultimately passed the Genetic Information Nondiscrimination Act in 2008 with overwhelming bipartisan support, which President Bush signed that spring; it was the first antidiscrimination legislation passed in the United States that was designed to preemptively prevent a future source of discrimination, rather than curtail a known history of past discrimination.

Collins was particularly adept at making genomic initiatives personally attractive to influential politicians. When he made the pitch to George W. Bush in 2003, the nationwide study was packaged as a father-son affair: George H. W. Bush launched the Human Genome Project, and George W. Bush would bring the medical applications to America. His "Leavitt-friendly proposal" from 2005, of course, was all about getting a Utah hook in the American Gene Environment Study so as to woo the former governor.

In August 2008, Collins stepped down from his role as director of the Genome Institute after leading it for more than fifteen years. He'd been envisioning writing a book on personalized medicine, and he sought the necessary time and mental space for it. He also wanted to pursue opportunities, he told journalists, that were not permitted of federal scientists. Collins, for instance, wanted to consult lawmakers on matters of national biomedical research. The chance to do that work presented itself quickly. Just after the presidential election, as Alexander was preparing to approach Kington with the menu of options for the National Children's Study, Collins joined the Obama-Biden transition team, advising on issues related to the Department of Health and Human Services.

. . .

Alexander returned from his meeting with Kington looking shell-shocked. The acting director, Alexander told his staff, went ballistic when he brought up the National Children's Study budget, accusing Child Health of hiding the true costs. The initial feeling at the institute was bewilderment. Why would presenting various funding-level models elicit such rage? There must have been some miscommunication. Alexander, along with Scheidt, visited Kington a second time to explain the history of the budget, how he navigated the budget directive, and what options he had for moving forward, hoping that conversation would clear up any misunderstanding. It did not. Kington accused Alexander and Scheidt of taking a $7 billion study and misrepresenting it for years as costing only $3 billion. That, he said, could only be the result of "deception or incompetence." The budget directive piece of the story was dismissed as a suspicious alibi; proving such a defense would require documentation. Kington ordered his Office of Management Assessment to investigate. That unit is like the NIH's department of internal affairs; it's located within the director's office and tasked with overseeing regulations, inspecting charges of fraud or waste, and recommending any corrective actions that need to be taken if problems are found. The investigation, Kington told Alexander and Scheidt, would involve a deep dive into all things related to the National Children's Study finances over the life of the project. After the meeting, as the two pediatricians walked back toward Child Health, Alexander glanced over at Scheidt. "That was the worst meeting of my life."

The Child Health staff spent the early months of 2009 responding to all the new demands of the investigation—turning over emails, finding and sharing budget documents from the previous nine years, participating in interviews. On the evening of Monday, March 16, the phone rang in Scheidt's home. It was Alexander. Kington, Alexander said, wanted Scheidt removed from his leadership position; somebody needed to fall on their sword, and the study director was the natural person. Alexander proposed having Scheidt shift over and lead Child Health's Center for Research on Mothers and Children. The National Children's Study director talked it over with his wife. The position that Alexander offered him was a respectable one. Plus, as his boss explained, it offered Scheidt a chance to leave quietly, before the Office of Management Assessment report was issued; there was no way to know for certain what would come out of it. On the other hand, Scheidt didn't feel that he or anybody else in Child Health had done anything wrong. They'd been reporting the $3 billion budget since 2005 because that's what they were told to do, and so he figured the most likely result of the investigation, no matter how probing, would be simply to confirm what they'd been saying all along. They'd already been talking openly with the National Children's Study site directors and advisory committees about the fact that the study, if confined to that total number, would need to be reeled in because of cost limitations. Stepping aside now, the fierce competitor thought, would be the equivalent of quitting. Alexander could have asked for Scheidt's resignation then and there. Instead, he let Scheidt make the choice. When his colleague decided to stay and fight, Alexander backed him and assumed any consequences of his characteristic loyalty.

NIH personnel, ten days later, appeared before the House Appropriations Committee. At the front of the spacious room, a long, wooden dais hosted two dozen chairs for members of Congress. At the back, several hundred seats were in place for reporters, lobbyists, and other advocacy representatives. Sandwiched between those audiences was a small desk with just two places to sit, one for Kington and the other for whoever else was asked to speak. The hearing was to discuss the general topic of the agency's budget for the coming year, but two topics were of particular concern. The first was the American Recovery and Reinvestment Act passed in February. That bill allocated roughly $800 billion for stimulus efforts designed to combat the Great Recession, more than $10 billion of which went to the NIH. Representatives wanted to hear about what Kington had planned for those federal dollars. The other topic was the National Children's Study. The acting director notified Congress about "adjustments and mid-course corrections" that he was making to the study due to the budget number shifting from $3 billion to $7 billion. "Earlier this year," Kington told the politicians in his prepared remarks, "we learned that, if all the potential components of the study were to actually go forward, the most recent total estimate would not be within the spending range expected by the agency."

The conversation surrounding the study played out differently for the politicians depending on whose microphone was turned on. For Jesse Jackson Jr., Democrat from Illinois, the National Children's Study was overdue for congressional attention "because President Bush tried his hardest to eliminate it." For the Kansas Republican Todd Tiahrt, Kington's news about the study offered a cautionary tale about the federal government's tendency toward mismanaging taxpayer dollars. A study originally budgeted at $3 billion, he understood, might actually cost more than double that. So the conservative politician was worried that the $10 billion they were giving the NIH as part of the stimulus plan would only translate into an ask for $20 billion down the road.

The Democrat Lucille Roybal-Allard from California, among the strongest supporters of the project in Congress over the years, was the one representative who, rather than sharing her personal view, wanted to hear the Child Health Institute perspective. "Is Dr. Alexander here?" she asked, looking out across the cavernous room. As Alexander stood up and moved toward the front, she asked him to respond to the accusations about the total projected cost. "Do you share that concern, in terms of doubling the budget?"

Alexander took his seat before the crowded room. He'd sat before members of Congress many times over the course of his nearly quarter century as director of Child Health, but those previous occasions were generally amicable affairs where he updated them on the progress being made combating sudden infant death syndrome or new treatments being tested for autism. This unnerving situation was nothing like that. Alexander had tried telling Kington how the budget took on the form that it did, but the acting director didn't buy it. The Office of Management Assessment investigation was still not complete, and yet Congress was informed that the budget was in disarray. Now Alexander had to defend himself and the study before Congress and the media while sitting next to the very person who ordered the pending investigation.

Alexander fidgeted with his jacket sleeve, rocked back and forth in his chair, and stumbled over his words. He admitted that he had indeed welcomed ambitious proposals about what sort of research could be conducted with the first nationwide cohort of children in four decades. But that was because the scientists simply didn't know which tests and experiments would perform well during the Vanguards' pilot testing and which wouldn't. It was also because he thought the study lent itself to collaborations with nongovernment entities like nonprofits and advocacy groups, which might be interested in funding research components above and beyond the federal investment. There were good reasons to encourage bold visions for what could be achieved, he said, even if that potential list of activities never became an actual reality. "There was never any anticipation that we would double the size of the study or even massively increase it," he assured the congresswoman.

After the hearing, the science reporter Meredith Wadman from *Nature,* who'd been listening from the gallery, pulled Alexander aside and invited him to say more about the claim that the National Children's Study budget had doubled. That was a "myth," he told her. That was not the accurate characterization of the circumstances surrounding its design and financing. The pilot testing phase presented to the scientific community was "an attractive tree to hang ornaments on," Alexander analogized; the Vanguards would then determine which ornaments made the cut. "There was never any expectation or intent that we would be able to fund all that."

While Alexander was defending the study in Washington, D.C., staff with the Vanguards were going door-to-door in the designated neighborhoods of New York, North Carolina, Utah, Pennsylvania, California, Wisconsin, South Dakota, and Minnesota trying to get families signed up for the study. The groundwork for the door knocking was laid with advertising campaigns in the communities—billboards featuring smiling children, television spots with pregnant mothers saying, "I can help find answers to autism." That didn't mean everybody in those neighborhoods was prepared for the visits, though. Some families didn't know what the recruiters were talking about. Others assumed they were federal representatives gearing up for the 2010 census. A recruiter in South Dakota came upon a gentleman barbecuing in his backyard naked.

The directors reported a variety of challenges in those early months. Recruitment was difficult during the winter because there was so little daylight, and inviting parents to enroll their child in federal science didn't seem like a fruitful conversation to have after dark. The handheld tablets that the recruiters used for collecting data about potential participants were also glitchy; the first iPad, after all, didn't hit markets until the following year, so the technology was new. The screens were hard to read in bright sunlight. They had poor battery life, and so sessions with families often had to be cut short because the devices went dead. There were also early signs of regional differences in how successful recruitment might be. In the counties that were fairly homogeneous in terms of racial and ethnic makeup, like Salt Lake, recruiters routinely hit their target numbers. But in the counties with a great deal of racial and ethnic diversity, like Queens, New York, the recruiters struggled. Part of it was a language barrier. Landrigan believed there were more than one hundred languages spoken in the Queens neighborhoods randomly selected for the probability sample. Even when the recruiters could communicate with potential participants, they often encountered a great deal of hesitancy to sign up because the immigration status of many residents in the region was vulnerable, making them uneasy sharing personal details with strangers who represented the federal government.

The problems encountered at the Vanguard Centers were frustrating but not devastating. Challenges were to be expected after all. The key was giving the Vanguard Center directors the time and space to figure out how everything that had been imagined on paper over the previous nine years translated onto the doorsteps of American families.

But over the course of three days in the summer of 2009, the fate of the National Children's Study changed forever. President Obama, on July 8, announced his nomination for the next leader of the NIH. He was traveling to Italy for a G8 summit meeting, so there was no press event, but the White House did issue a statement. "The National Institutes of Health stands as a model when it comes to science and research," the president declared, adding, "My administration is committed to promoting scientific integrity and pioneering scientific research and I am confident that Dr. Francis Collins will lead the NIH to achieve these goals."

On July 10, the Office of Management Assessment released "Internal Control Review of the National Children's Study." The National

Children's Study leadership, it judged, routinely provided inaccurate numbers for the total projected costs of the study, omitting anticipated direct costs, indirect costs, and inflation. The most damning conclusion of the report, though, was that the mismatch between the lower estimate and the higher estimate wasn't simply a point of confusion or error of oversight. Rather, the investigators pointed to Keim's emails as evidence of something more devious. In September 2006, recall, Keim told a Child Health budget officer, "To date, we haven't been incorporating inflation into the numbers in order to maintain the feeling that the total cost of the Study is within some realm of fund-ability. I suspect that our strategy on that front has not changed, and I also suspect that if the total cost of the Study suddenly becomes something far above what we have been sticking to that Bldg 1 may react negatively." Then, in May 2007, remember that Keim wrote another budget officer, "Unfortunately, we are all sticking to some oversimplified assumptions about the cost of the Study (like that it will cost $99.4m per year for the last approx 20 years of the study)." For the Child Health personnel, the statements were referencing the Bush-era budget directive. In 2009, however, they could locate no hard evidence of that directive: no email, no letter, no meeting minutes. Keim's two emails, to the Office of Management Assessment investigators, were instead the smoking gun revealing that "cost estimates were consciously maintained consistent with previously reported levels to avoid negative reactions from the NIH [Office of the Director]." The same day that the report was issued Alexander went to Scheidt's office. Leaving the study now was no longer optional. Alexander hand delivered the letter that officially removed his longtime friend and colleague from directorship of the National Children's Study.

The report was supposed to be confidential, but Scheidt's removal and rumors of the accusations reached the media. "More Turmoil over Hidden Costs of NIH Children's Study," read a prominent headline in the magazine *Science*. The budget story did more than just threaten those in charge of it. With a global recession unfolding and talk of the total cost projection doubling, funding any study at all was now a point of discussion.

The Interagency Coordinating Committee, which advised Alexander on the National Children's Study, met in August, with news of the investigation results and the negative publicity swirling across the federal government. A dozen representatives from the NIH, the CDC, and the EPA gathered in a cramped conference room at the EPA's downtown D.C. headquarters. Alexander occasionally attended a portion of those events, although it typically involved him just phoning in to the group. He attended in person only when there was something of great importance to share. The room was somber when Alexander arrived; the usual chitchat that fills a space before a meeting is called to order was replaced with the sound of chairs squeaking and papers shuffling. "This is a turbulent time, to put it mildly," Alexander began. The Child Health director was calm and composed, but he spoke slowly, with no sign of the cheerful, optimistic pediatrician the group had come to know and respect.

"Let me set the record straight one more time. I've done this with you before, but I think it's probably worth reinforcing. We are accused of, basically, three things." First, intentionally leaving the indirect costs of the study out of budget estimates. Second, failing to notify anybody in NIH, Health and Human Services, or Congress that the costs had increased. And third, reporting a total projected cost that knowingly concealed the actual projected cost so as to misleadingly maintain enthusiasm for the study. "None of these charges are true," Alexander said.

Leaving the indirect costs out, Alexander admitted, was an accident. Those should have been built into the original budget, and they were not. But that was a mistake, a mistake missed for several years by the Child Health budget office, the NIH budget office, and the Health and Human Services budget office. It was not a deliberate misrepresentation on the part of Child Health. Indeed, it was an error that Alexander took to the NIH leadership and tried to correct. "The response from the Office of Budget people at NIH," Alexander reminded the room, "was, 'You can't do that!'" He intended to go back and try correcting it again in 2006, but then the call for terminating the study came along, which persisted for the remaining years of the Bush administration. "And so we were stuck."

Things went sideways when Alexander had his meeting with Kington after the election. "I remember it well," Alexander mumbled, his voice now starting to break. "I wish I didn't." Kington's initial reaction—that the whole fiasco was the result of deliberate duplicity on the part of Alexander and his staff—became the official version of events that was conveyed to the upper levels of the federal government. The Office of Management Assessment report, which he hoped would exonerate them, instead came back with a guilty verdict because they lacked the documentation required to prove that they were responding to the budget directive all along. "We are trying," Alexander continued, "to provide information to the members [of Congress] about the 'other side of the story,' as we call it. The fact that we are not guilty as charged."

That was a challenge, though, because Alexander, after more than forty years in federal service, was being shunned as a distrustful disgrace. He met with Collins and explained everything, but Alexander said the new NIH director was unmoved; on the contrary, in fact, Collins told the Child Health director that he was getting pressure to have Alexander join Scheidt out the door. The next person in the chain of command would be Obama's new secretary of health and human services, Kathleen Sebelius, but "the well has been poisoned in the office of the secretary so effectively that nobody seems to be able to make a dent in it or get through. The secretary is saying things like, 'They cooked the books and lied to the Congress when asked about the study.'" Alexander, after pausing for a moment, added, "That's hard to overcome." Congress, for its part, described the whole scenario after the investigation report as a "breach of trust," meaning even their biggest allies in the House and Senate had little room to operate.

After about twenty minutes, Alexander took a break, sounding exhausted, and invited questions. The Interagency Coordinating Committee members voiced how difficult was the bind in which they found themselves. On the one hand, they wanted to defend Alexander and Scheidt from the miscarriage of justice that was unfolding. On the other, they wanted to salvage the National Children's Study, which risked being shut down entirely as a result of the media attention and political exposure it was receiving. Go to bat for their two colleagues, and they could undermine faith in the study further. Rally around the project, and they cut Alexander and Scheidt loose. For the Child

Health director, the choice was clear. He was no longer in a position to advocate for the National Children's Study, but the others in the room were. They should speak to the heads of the CDC and EPA and encourage them to publicly express their ongoing commitment.

The study was housed in the NIH, though, so the person who now had the most control over it was its new director. Where did he stand? "It's hard to say completely," Alexander replied. "I think he's supportive of the study in its concept." What exactly that concept was in the mind of Collins remained to be seen. Alexander told the group that Collins held his first town hall with NIH employees just a few days earlier, and he brought up the National Children's Study in response to a question about NIH research resources available to address racial health disparities in the United States, but then Collins immediately pivoted to his desire to conduct an even larger study, involving genes and the environment, with 500,000 Americans of all ages. "But this is not the first time this has surfaced. Back in 2003 . . ." Alexander began, and then he got cut off by a chorus, "We know about AGES!" The moment offered a brief, collective chuckle. "It is still alive in Francis's mind," Alexander went on after the laughter subsided, and so it wasn't clear what form the study might take under Collins's leadership.

"Is there anything that this body can do for you?" Alexander was asked. He took a long time before answering that offer of support. "Um . . . so far . . . NIH has controlled all the information that the secretary receives. And they're so convinced that there are some really bad guys here who deserve to be punished for keeping all this information secret, and not disclosing it, and not sharing it, and covering up—that's the only message that she has heard, or her senior staff has heard."

With time running out, one of the committee members shared his experience on a recent conference call with National Children's Study affiliates like Landrigan and Edward Clark. "There was outrage," he reported. The faith in Alexander and Scheidt was universal, along with the disgust in how they were being unfairly shamed and sacrificed.

The Child Health director exhaled, and then he hoarsely whispered, "This is not the way the government we work for is supposed to work." Alexander was visibly emotional now, and he wasn't alone. Others in the room bowed their heads and covered wet eyes with their hands. "We all made commitments of our professional lives and careers to this government and the democratic process. And it's been *violated badly,*" the last two words uttered in seething agony. He gathered himself and slid back from the table. "I better go."

Alexander mustered his pleasant demeanor for Child Health's regularly scheduled advisory council meeting just a few weeks later. They were trying out an early version of videoconferencing at the gathering, and the institute director was both patient and enthusiastic with the handful of members joining remotely. In the room itself, ten to fifteen advisers sat around an enormous, oblong conference table, while dozens of staff and guests watched from chairs about the periphery. Several representatives from academic societies were also present in the room. "Hooray!" Alexander shouted and waved to them in the back of the room.

Among the first orders of business were personnel changes at the institute. Alexander, without mentioning the Office of Management Assessment investigation or report, informed them that Scheidt was no longer overseeing the National Children's Study. Alexander had asked Steven Hirschfeld to step in and serve as the acting director. Hirschfeld at the time was officially the associate director of clinical activities at Child Health, but Child Health had little in the way of clinical activities, so instead he focused on bookkeeping activities like getting medical terminology across the institute harmonized and creating new forms. This amounted to issuing memos to various institute programs, asking them to change the way they were doing something or wording something or documenting something to align with his new protocol. Child Health employees, for their part, tended to find Hirschfeld annoying and so just ignored him. But it wasn't only his odd collection of seemingly irrelevant activities at the institute that made him an unusual choice to run the study; he also had little experience with managing large, longitudinal studies. There were other federal scientists at the institute who did, but they were busy with their own research and administrative duties in 2009. Hirschfeld's marginal responsibilities, it turned out, made him a suitable fit. Alexander, when Scheidt left the study in July, was just looking for someone to keep the seat warm until a search for a new director was completed. Hirschfeld could be temporarily moved into the position without disrupting significant activities at the institute.

That search for a new study director, however, would proceed without Alexander overseeing it. After introducing Hirschfeld, Alexander added that there was one more personnel change "not on your list because of late breaking developments." He was going to be stepping down as head of Child Health in a matter of days. Just as Alexander shared the sudden news, he looked up and added, "Dr. Francis Collins, who I was about to describe as the new director of the National Institutes of Health, has arrived." Collins walked briskly to the front of the room, took his seat next to Alexander, smiled broadly, and thanked the council members for their service, emphasizing how important the Child Health Institute was to the NIH and how committed he was to it in this time of leadership transition. He then turned his attention to Alexander, noting, "It seems like an appropriate moment to point out some of the main accomplishments that we can all point to, and celebrate Duane's time here." Collins began by heaping praise on the pediatrician's support for research that showed prenatal genetic testing was safe and effective and then turned to other highlights of Alexander's career, such as Child Health's role in combating sudden infant death syndrome with the Back to Sleep campaign. The NIH director concluded by encouraging the room to give Alexander a polite round of applause and then got caught off guard when it turned into a thirty-second standing ovation.

Collins wrapped up the show of appreciation by giving Alexander a few friendly pats on the shoulder, but the whole affair had an air of discomfort. Pleasantries were quick; smiles were brief; eye contact was minimal. Alexander rarely looked up from his folded hands during Collins's monologue. "I'm not able to take questions this morning," Collins told the audience. For the pediatrician who'd spent more than two decades as director, leading the institute for more than half of its existence, it was an unceremonious end to a distinguished career. No public celebration, no warm tributes from staff and colleagues, no photos with national politicians. One observer described it as more like a scene from *The Godfather*—an executioner singing the praises of the soon-to-be departed.

It was for Collins now to pick the new director of Child Health. He briefly appointed a deputy director from the National Heart, Lung, and Blood Institute to come over and serve as acting director, but she quickly returned to that institute to become acting director there. So Collins turned to his longtime friend, trained pediatrician, and deputy from the Genome Institute to fill the post: Alan Guttmacher was going to lead the Child Health Institute.

In November 2008, Alexander was so emboldened by the favorable National Academies review of the National Children's Study, the impending arrival of the Vanguard Centers recruitment, and the election of a Democratic president that he made the case to Kington for what could be achieved with a larger study budget. By November 2009, just one year later, Alexander along with Scheidt was gone; the study was shrouded in accusations of deceit and mismanagement; and the whole project was housed in a federal agency overseen by the person who'd had the most contentious relationship with it.

That same fall, Jocelyn Kaiser, a reporter with *Science,* interviewed Collins about his new leadership position and what it held for the future of medical research in America. How would the Great Recession impact federal science? Was the new director excited about President Obama's intention to double investment in cancer research over the coming eight years? Her last question was about the National Children's Study and its troubled budget. Should the full study go forward? Collins was committed to the science, he replied. The questions were, what could be afforded, and what was going on with the pilot testing? "We're going to take a very serious look at the study design," he told her.

6

Genomics Finds Its Hammer

IN 2007, THE LEADERSHIP at NIH's Child Health Institute invited the National Academies of Sciences, Engineering, and Medicine to come and provide an independent assessment of the National Children's Study. In 2013, it was Congress demanding the review, prohibiting the funding of new study-related contracts until that evaluation took place. They sought the National Academies' services because an insurrection that started out in private grew increasingly public and increasingly vitriolic. At the heart of it, researchers and advisers to the National Children's Study feared that the scientific effort had become unmoored, charging the new leadership at NIH with abandoning the original intent of the project.

The National Children's Study, by 2013, had become a casualty of what the philosopher Abraham Kaplan, fifty years earlier, called the "law of the instrument." Kaplan was among the more prominent public intellectuals of the 1960s, appearing on the cover of *Time* magazine and in documentaries about inspirational educators. He had a knack for seeing philosophical problems that cut across disciplines. The law of the instrument was one such observation, which Kaplan saw unfold in fields ranging from quantitative educational research to the science of library categorization. "Give a boy a hammer," the law goes, "and everything he meets has to be pounded."

Kaplan thought the law of the instrument was a natural human inclination. When a new technology came along, be it quantitative tools in the behavioral sciences or automation in libraries, enthusiasts tended to think it solved everything. Rather than adapting new tools to old jobs, the advocates tried to morph the old jobs to fit their new tools.

Geneticists, by 2010, had their hammer. The swift drop in the costs and time of genotyping and DNA sequencing made those high-throughput technologies suddenly available to geneticists with all sorts of curiosities. Companies like Illumina, we've seen, explicitly developed the machines and resources that made genotyping and sequencing widely available because they knew they were necessary to make pharmacogenomics, the lucrative heart of personalized medicine, a reality. But once those costs dipped significantly, scientists with no interest in drugs could use them for their own purposes. A bit of bone from a Neanderthal, a species on the verge of extinction, a child heading off to kindergarten—anything with a genome became a viable candidate for providing some DNA to a geneticist.

A pattern quickly unfolded in disciplines as disparate as archaeology and educational policy. Geneticists, emboldened by their new big-data tools, swooped into fields with promises of revolution, of innovation, of answering the most important questions. In reality, the DNA-minded scientists bent those disciplines toward the sorts of problems that their genomic technologies could help address. What unfolded at the National Children's Study wasn't unique. It was just a particularly expensive and publicized example of what happened when scientists in one of those domains revolted against being hammered.

. . .

Over the course of a couple of months in 2010 the field of archaeology was rocked with back-to-back-to-back publications. In February, a lock of hair from a man who lived four thousand years ago among the Saqqaq settlers of Greenland was used by researchers from the Natural History Museum in Denmark to provide the first whole genome sequence of an ancient human—"Inuk," the team named him. An international team led by scientists from the Max Planck Institute for

Evolutionary Anthropology followed a month later with the report of a new, previously unknown hominin distinct from both modern humans and Neanderthals based upon their analysis of mitochondrial DNA from a female fingertip bone found in Siberia's Denisova Cave; she received the name "X-woman" because mitochondrial DNA is passed down through generations on mothers' X chromosomes. Then May brought the draft sequence of the Neanderthal genome, derived from samples that belonged to several prehistoric individuals.

This wasn't the first time that genetic information was brought to bear on human evolutionary history; in 1987, snippets of mitochondrial DNA from almost 150 people spread out across the globe were used to argue famously that all modern humans were descendants of a "mitochondrial Eve" who lived in Africa approximately 200,000 years ago. The 2010 papers, though, signaled something very new. Previous genetic research that allowed claims about prehistoric humans typically relied on a little DNA drawn from people living in the present; that was then used to make inferences about humans living in the past. These new studies, instead, took advantage of Illumina's latest platforms to probe enormous stretches of DNA from the ancient humans themselves.

Media reports were smitten with what the troves of genetic data revealed. The studies themselves were published in the premier scientific journals such as *Science* and *Nature*. Ancient DNA (sometimes abbreviated aDNA, or referred to as paleogenetics or archaeogenomics) was said to "rewrite human history." Beings from the past were genetically resurrected with actual names and artistic re-creations depicting shaggy hair and piercing stares. National Public Radio declared 2010 a "Good Year for Neanderthals (and DNA)." Of all the places where genomic technologies migrated out of medicine, human history was a natural place for geneticists to make a quick and newsy impact. With just a bit of bone or hair or tooth delivered by archaeologists eager to add their names to a paper published in *Nature*, geneticists had the opportunity to make international headlines. Stories of newly discovered humans, of our Neanderthal-cuddling ancestors, and of one ancient group dramatically wiping out another suggested that the genomic revolution that had started in health care could now answer archaeology's biggest questions.

Journalistic excitement built off claims the geneticists made in their publications and to the press. David Reich, a population geneticist at Harvard Medical School, was among the most outspoken advocates of this "ancient DNA revolution." Reich contributed to the Neanderthal genome draft and then also participated in efforts to scan the nuclear genome of the Denisova Cave sample, describing X-woman as belonging to a newly designated hominin group, the "Denisovans." He subsequently set up a lab in Boston devoted to research on ancient DNA, helping drive a series of high-profile reports: the full sequence of the Neanderthal genome later in 2013, then an analysis of genomic samples from humans living thousands of years ago across what is now central Europe, Ukraine, and Russia indicating a process whereby herders from the Eurasian steppe swept in and displaced the Neolithic peoples occupying Europe at the time. Reich equated his focus of the genomic technologies on ancient remains with Leeuwenhoek's seventeenth-century discovery of strange microbes in pond water using the light microscope, a historic event filled with scientific surprises. The result was a "gold rush," according to Reich; ancient human remains discovered at dig sites were like "gold nuggets strewn on the ground" awaiting the attention and expertise of geneticists.

Reich and other investigators of ancient DNA, with genetic information derived from samples that came from a handful of archaeological sites, offered up sweeping theories of human history. Massive migrations took place on a continental scale. Farming cultures from the East stormed in and replaced hunter-gatherers from the West. Human evolution looked like a *Game of Thrones* season, with new groups conquering old and then finding in time that they too were threatened by a sudden arrival. Reich, for his part, extended this dramatic portrayal to the ancient DNA revolution itself, proudly comparing geneticists to barbarians at the gates of professional archaeology. "We haven't gone through graduate school in anthropology, or linguistics, or history," he granted, "and yet we're making very strong statements about these people's fields. It's a little bit like barbarians are walking into your room, and you can't ignore [the] barbarians because they have information, weapons, and technology that you didn't have access to before." The science of human history was about the big questions of migratory movements and interbreeding relationships according to the population geneticists, and so they felt they were in the unique position of being able to provide the big answers.

Archaeologists, for their part, had seen claims of massive migration and conquest before in their field; much of nineteenth- and early twentieth-century archaeology was rooted in such themes, and the sudden return was unsettling. The scientific narrative was about the spread of civilization across continents, one culture replacing another over a short period of time. The German linguist and archaeologist Gustaf Kossinna embodied those ideas and popularized them in European archaeology in the first decades of the twentieth century. He professed a theory of "settlement archaeology" whereby archaeological finds that reflected shared material culture—a certain style of pottery, or a burial practice—indicated shared ethnicity. Kossinna's particular interest was in using artifacts to trace the history of his contemporary Germans back through their ancestral past. What they uncovered, according to Kossinna, was a story about the rise and civilizing virtues of the German people who historically dominated a homeland far beyond the boundaries of what were then Germany's internationally recognized borders.

Nazi leaders eagerly embraced Kossinna's brand of archaeology. The science was weaponized for political purposes, used as a tool of propaganda designed to unite a population and tie their struggle to a deeper history. It was also used to justify an expansionist agenda; if Germany's cultural ancestors were located in Poland, as Kossinna proclaimed based on his interpretations of artifacts found there, then the Nazis saw themselves as having every right to reclaim that region as their own. The image of superior ethnic groups conquering inferior ones across human history found favor with a totalitarianism that fed off nationalist instincts.

Starting in the 1960s, archaeologists made an intentional move away from the Kossinna style of theorizing. That was partly about breaking from a vision of archaeology that was so easily politicized, and it was also about doing science that was better attuned to empirical nuance. Closer inspection of the groups historically lumped together as unified, ethnic wholes revealed a great deal of complex cultural heterogeneity across different regions and different times. "Pots are not people," went the corrective mantra. This "New Archaeology" or "processual archaeology" shifted close attention to the manner in which human groups responded in very specific ways to the unique, local environmental challenges that they faced where they lived. Episodes from the past that were traditionally characterized as sudden clashes of culture were found to have played out over much longer periods of time and to have involved complicated processes whereby groups mingled with one another, exchanging ideas and resources.

The theoretical reorientation within archaeology in the late twentieth century brought with it a different way of practicing that science too. Attention shifted to local communities and cultures, where archaeologists could spend an entire career meticulously examining a prehistoric site. Doing so allowed them to look for clues that shed light on questions about how those particular people lived in that place, what they ate, how they dressed, with whom they traded, how they died, why different people within the group were buried in different ways, which members of the group played which roles, whether the society was organized hierarchically or in an egalitarian fashion. That long-term and localized investment in a specific place also permitted archaeologists the opportunity to form relationships with communities that continued to occupy the same space or viewed themselves as descendants of the ancient people, facilitating a level of trust and cooperation that was a departure from archaeology's past, sometimes parasitic, interactions with Indigenous communities.

The ancient DNA revolution as espoused by population geneticists like Reich morphed archaeology away from those fine-grained analyses and back in the direction of Kossinna's preferred scale. It was certainly not an intentional effort to make contemporary archaeology more conducive to Nazi sympathies. But it was an instance of the law of the instrument. The geneticists didn't redirect attention to massive migrations and hereditary relationships because some independent scientific body voted those topics to be archaeology's "most important"; instead, they focused upon those topics because they were the ones that their genomic technologies could hammer with the few samples that they had, and they then dubbed their nails the "most important." The actual science of ancient DNA studies was an extension of the larger field of comparative genomics, which used genetic information about different individuals or different species to assess questions of relatedness. It made estimations of how far back in evolutionary history you and a chimpanzee or even a pumpkin had a common ancestor. The geneticists who turned this science and technology on ancient human remains similarly made judgments of genetic affinity and difference. The fact that modern humans and Neanderthals shared roughly 2 percent of their genetic material, for example, pointed to interbreeding between Neanderthals and our ancestors in the distant past. The Denisovan genome was significantly different from both the Neanderthal genome and the modern human genome, which the population geneticists used to say it was a distinct hominin. Migration inferences also followed from the results of comparative genomics. The Danish scientists who sequenced Inuk in Greenland discovered that he was more closely related to the Chukchi and Koryak people who currently live in Siberia than he was to the modern-day Inuit of Greenland; that told the researchers that the Saqqaq who colonized Greenland originally came from northeast Asia but were then eclipsed in some way by the ancestors of the modern Inuit.

Those revelations were genuinely interesting, but that didn't make them archaeology's biggest questions. Inuk's DNA couldn't tell scientists about the social dynamics of the Saqqaq, or how toolmaking practices were transmitted, or what they fabricated to survive the harsh climate; only close archaeological examinations of the artifacts from that region would reveal those phenomena. Inuk's DNA could indicate that a migration took place, but it couldn't say why generations of Inuk's own ancestors made that trip or deemed Greenland a suitable final destination. What's more, the genetic affinity between Inuk and modern human groups in Siberia said nothing about what became of the Saqqaq.

The clearest way to delineate the limits of research on ancient DNA is to reflect on what we know, or rather what we don't know, about the Denisovans. The purported discovery of a new species of humans was hailed as a historic scientific finding. And yet, almost everything that science can currently say about them comes from the DNA of just a few teeth and bone fragments. They are genetically different from Neanderthals and our modern human ancestors but interbred with both; they occupied the Denisova Cave between roughly 50,000 and 200,000 years ago; a number of modern human groups across Asia and the South Pacific retain the greatest genetic affinity to them. Beyond that, the Denisovans as a people are a mystery precisely because so little is known beyond the genetics. We do not know how they lived, how they interacted with one another, or the details of their encounters with the other hominin groups in the area. Little can be confidently said about Denisovan culture because, while many artifacts have been found in the Denisova Cave, that space was historically occupied by Neanderthals, Denisovans, and early modern humans, making specific artifact-Denisovan links an ongoing source of debate. In fact, they don't even have a proper scientific name yet. The Denisovans are counted among the genus *Homo,* but they still don't have an official species classification (like *sapiens* or *neanderthalensis*), because the scarcity of physical specimens makes their taxonomic status murky.

Archaeologists were not opposed to the idea that genetic data could provide valuable information for their science. Indeed, they celebrated occasions where geneticists worked closely with archaeologists, using their novel technologies in ways that archaeologists knew that science could be most insightful. The Lech River valley south of Augsburg in Bavaria provided one striking example. Over a narrow strip of land just ten miles long and wedged between the Lech and the Wertach Rivers, archaeologists uncovered hundreds of burial sites associated with a number of different groups that lived in the region thirty-five hundred to forty-five hundred years ago, spanning the transition from the Neolithic to the Bronze Age in central Europe. Geneticists then provided DNA analyses of eighty-four individuals who lived among those different groups. What they discovered was a patrilocal residential arrangement, where male leaders stayed put and adult females moved in from other communities. This revealed a process whereby cultural transmission between groups could play out at the very specific and very local level of individual women carrying ideas and resources with them as they left the home where they grew up.

Archaeologists were critical of a certain style of genetics that made grand inferences from just a dozen sites distributed across continents. They bemoaned the return of a blunt ontology that lumped together enormous populations spread out over vast time and space as "farmers" and "hunter-gatherers." They disagreed with the inclination to treat migrations as black boxes that just happened rather than as structured behaviors involving actual people. They challenged the notion that the geneticists' tools were somehow intrinsically more scientific and reminded the biologists that their DNA-based methods relied heavily on the actual work and expertise of the archaeologists to find, unearth, date, and assess the context of every sample that the geneticists studied. Archaeologists were also worried about a return to the predatory practices of the past. If analyzable DNA is to be recovered from ancient remains, portions of the bone or teeth must be destroyed so that the material can be ground up for DNA extraction. There was good reason to be concerned; Indigenous populations, frustrated by the way ancestral body parts were being disinterred and damaged without their input, labeled the ancient DNA research "vampire science" and sought to shut it down, threatening archaeological access to sites.

What worried the archaeologists most, in short, wasn't ancient DNA; it was the way that a purported DNA revolution ignored all the scientific and moral progress that archaeology had made over the previous five decades. This genomic imperialism brazenly moved the discipline back toward epistemologically and ethically shaky ground. Reich's embrace of his "barbarian" status was born of technological hubris. It was but one instance from that period where geneticists ignored a discipline's history and tried to mold it in such a way that they and their particular tools were paramount.

. . .

Imagine a new school opens up in your neighborhood. It sits on a sprawling campus, large enough to accommodate the younger grades of elementary school alongside the upper grades of middle and high school; there's also an integrated center for children with special educational needs. A thriving arts program takes advantage of the music rooms, media rooms, and theater. For the more biologically inclined pupils, a horticulture center offers botany and beekeeping. The sporting facilities are state-of-the-art too, with a pool, manicured athletic fields, and a fitness center. Small classes allow for ideal learning and socializing environments. Books can be checked out from the school's large, well-stocked library. The teachers have total control over the curricular material that they develop for their students. When it's time for physical education, the students can opt from any number of different activities—fencing, archery, gymnastics, yoga. The administration, for its part, is constantly testing out new, innovative pedagogical interventions; that way, all the methods utilized in the school are evidence based. The children's physical, mental, and emotional health is also closely monitored; pediatricians, nurses, speech and language therapists, and counselors all work there.

If you have a child—let's call her Maria—you're probably wondering if you can get her into this outstanding educational system. Luckily, there isn't a complicated admissions process; no lotteries; every child in the community is accepted. When Maria is enrolled, she will be assigned a trained educational psychologist who will serve as her aide. That aide, Maria, and her teacher will meet before the first day of school. This is so an individualized education program (IEP) can be developed for your daughter. In most schools, these IEPs are produced only for students with special needs, but this school is different. Every child at the school has one so that every child's education can be tailored to their abilities and interests. As Maria matures, her aide will stay connected, ensuring that the insights derived from her performance in elementary school can be communicated up through middle and high school.

There's one more thing, though, the most important thing really. This is a "genetically sensitive school." Before that first meeting with Maria's aide and teacher, her DNA will be processed on an Illumina-style "Learning Chip" that will identify Maria's biological strengths and weaknesses. Is she at risk of developing a learning disability? Would she benefit from some particular reading intervention? Is she more verbally or mathematically inclined? Will athletics come naturally? This way, Maria's IEP will be guided by her own genome.

This genetically sensitive school is described in Kathryn Asbury and Robert Plomin's *G Is for Genes,* a book that calls for utilizing the insights of genetics to reform a defunct educational system. After introducing the basics of DNA in early chapters and then explaining the breakthrough that arose with cheap and fast genotyping and sequencing, they point to efforts to deploy those technologies on nonmedical, cognitive traits; the final chapter is Asbury and Plomin's attempt to pull all that science together and design a school that implements their genetic lessons.

Plomin, one of the foremost figures in the field of behavioral genetics, spent decades looking for regions of the human genome associated with intelligence. Initial hope in the 1990s for common genes with major effects on brainpower eventually went the way of so much in human genetics when it became clear that those genes didn't exist for cardiovascular disease, let alone IQ. Behavioral geneticists then followed the lead of the medical geneticists in turning to genome-wide association studies. That too, however, proved frustrating. Just as the medical geneticists wrestled with the missing heritability associated with traits like asthma and Parkinson's disease in the early rounds of their small-scale studies, Plomin led a 2010 genome-wide association study that included almost 8,000 individuals and looked for single nucleotide polymorphisms linked to cognitive ability. The effort came up empty. So the behavioral geneticists followed their medical counterparts again and increased the number of people included in their genome-wide association studies from roughly 10,000 to more than 100,000, seeking out common genetic variants with even smaller effects than previously anticipated. That finally did the trick. In 2013, a large, multinational research team pulled together more than 125,000 samples from dozens of different cohorts, most of which were genotyped on Illumina machines, and found three single nucleotide polymorphisms associated with years of education, which the scientists used as a proxy for cognitive ability.

The SNPs found in that study, not surprisingly, accounted for a very small portion of the variation in education years. Still, after years of failed attempts to link up specific stretches of DNA with something resembling intelligence, any result at all was considered a success by the gene hunters. When combined, they could provide a polygenic score just like the one the medical geneticists produced for traits like coronary artery disease and diabetes. The excitement fed calls for "personalized education." Around that time, scientists from a number of disciplines took advantage of the growing enthusiasm surrounding personalized medicine and extended the DNA-based model to their own domains. Personalized nutrition, personalized skin-care products, even personalized matchmaking all promised to use an individual's genetic information to pair them with their ideal vitamin supplement, skin moisturizer, or soul mate. Most of those exercises were little more than marketing gimmicks advertised by short-lived companies, which was, in part, what inclined Senator Obama to propose regulating those activities in his Genomics and Personalized Medicine Act. Personalized education, though, gained traction. *G Is for Genes* was just the first book, along with articles, op-ed pieces, and other volumes, that used the sudden affordability of DNA-based technologies to call for a genomic revolution in schools. Just as the advocates of personalized medicine criticized the cookie-cutter, one-pill-fits-all model of pharmaceuticals, geneticists like Plomin took aim at the "factory model of schooling," which supposedly treated all children as if they were identical blank slates. And just as the advocates of personalized medicine pointed to DNA as the solution to delivering individualized care, so too did the geneticists see "educational genomics" as the path to a school system that finally attended to the differences between children.

Personalized education enthusiasts would have you believe that tailoring education to students' aptitudes was a novel idea. It was not. This was yet another example of geneticists paying no heed to the history of the discipline they claimed to be revolutionizing. Throughout the mid- and into the late twentieth century, dividing students up based on their particular abilities was actually the norm. In elementary grades, the most common form was ability grouping where, for example, a fifth-grade teacher would break up his whole class into smaller groups for a particular subject like reading; the different groups, depending on their ability level, might then proceed at different paces, or be assigned different tasks, or read different texts. A more extreme version of this in elementary grades involved leveling, where entire classes were partitioned by performance, such that one fifth-grade teacher taught the highest-achieving students in all subjects while another fifth-grade teacher attended to the lower-performing students in a different classroom. While ability grouping and leveling were occurring in the lower grades, upper grades employed a system of tracking. Different subjects had different classes with different degrees of difficulty—Honors English or English, College Prep Biology or General Science, Algebra I or Algebra II; students were then sorted into the classes based upon their proficiency in those subjects.

Educational policy, like archaeology, underwent an evolution toward the end of the twentieth century. Critics throughout the 1970s and 1980s pointed out that the methods used to group students based on ability level also tended to group students based on race and socioeconomic status. It was no coincidence, they said, that middle- and upper-class white kids more often appeared in the higher-performing environments, while poor children and students of color occupied the lower-performing environments. Ability grouping, leveling, and tracking were justified on the grounds that they allowed students to maximize their unique potentials based on their exposures to performance-appropriate materials. The distribution of who wound up where, however, suggested a different process whereby inequities in society were reconstituted in the classroom.

These practices didn't just mirror societal inequities; they institutionalized them when finite educational resources were distributed to the different groups/levels/tracks. The students deemed performing lower generally received poorer resources, less qualified teachers, and less engaging lessons. Where students in College Prep Biology were outside investigating the ecology of a local estuary or inside dissecting a cat, students in General Science were reading the faded pages of an outdated book chapter about the scientific method. What's more, once students were assigned to a particular station, it was difficult to move. Teachers, administrators, parents, and even the students all knew whether any given child was in an upper or lower level regardless of what the groups were called, and so a self-fulfilling prophecy surrounded students in lower groups with lower expectations and students in higher groups with higher expectations. The result of all that was a stratified educational system that compounded inequality. It worked well for the students placed in upper-level programs where they were challenged, encouraged, and supported, but it left the lower-level students behind.

Throughout the 1990s, educational policy makers at local, state, and national levels shifted away from ability grouping, leveling, and tracking. Major educational and civil rights organizations like the National Education Association and the ACLU voiced opposition to the practices, and it impacted the actual work of teachers, who reported a significant decrease in these stratification efforts into the early years of the twenty-first century. That trend, however, gradually subsided, and the practices are on the rise again. Why this is the case remains a point of contention. Is it because of a newer generation of teachers without exposure to the earlier controversy? Is it because grouping students based on ability level is easier for overworked teachers? As the pendulum swings back in the direction of these methods, the current discussion among teachers, administrators, school boards, policy makers, and educational scholars revolves around whether such groupings can be done in a way that benefits all students and not just the well-off.

The educational policy debates that played out for decades were not about whether children were "blank slates." They were not about the virtues and vices of a "factory model of schooling." They were about trying to create an educational system that worked for different children with different abilities without worsening inequality. They were about the real-world challenges associated with doing right by all children with limited available means. Teachers didn't lack awareness regarding the value of personalizing education for each student; they lacked the resources—time, space, support staff, small class size, curricular materials—to do it fairly. The genetically sensitive school in *G Is for Genes* simply fantasized all those problems away. Asbury and Plomin admitted it was a utopian vision. But the very act of dreaming all the environmental inequality away was precisely how they morphed the educational policy conversation into a nail so that genomic technologies could go on hammering. Once the behavioral geneticists idealized an environment devoid of social, racial, economic, and educational inequity, personalizing education with "Learning Chips" was framed as the ideal way to let all children maximize their natural potential.

When we move from the utopia where the genetically sensitive school resides to the actual world replete with inequality, implementing educational reform guided by DNA isn't at all an obvious route to fairly maximizing children's potential. The most likely scenario is that the genomic interventions simply will not work; the genomic regions associated with things like educational attainment account for so little of the effect in question that making predictions about a child's potential and then designing interventions based on their DNA is a fool's game. But, just for the sake of argument, imagine for a moment a future where that personalized prediction does get some purchase. Utilizing genomic technologies requires financial and social resources, and therefore different racial and socioeconomic groups will have different levels of access. Wealthy, well-educated families will be more likely to hear about, understand, show interest in, and procure such genetic testing compared with lower-income and poorly educated families. Exclusive and expensive private schools will be able to offer amenities that allow different students with different aptitudes stimulating and challenging curricular paths, whereas financially strapped public schools will be limited in the extent to which they can do anything with such information about their students even if they have it. Concerns about self-fulfilling prophecies in the classroom will only magnify when the expectations about student performance are buttressed with information about their DNA. Genomically personalized education, far from paving a way toward a fairer world where all children can reach their potential, would more likely exacerbate inequity by combining unequal social and educational environments with disparate access to the latest science and technology.

. . .

"You may think of me as the Genome Guy," Francis Collins liked to quip to NIH employees after assuming his new position as head of the agency, "but I'm going to focus on all of NIH, and not just Genome." Collins's focus on all of NIH, however, took a very particular form once he became director. He identified many of the things that served the science of genetics well when he ran the Genome Institute and made them priorities for the whole agency. He encouraged, for example, the widespread use of high-throughput technologies so that institutes and centers across the NIH could find their own path to omification and get in on big-data science. He also wanted to see more public-private partnerships between the NIH's federal researchers and those from industry, much like the Pfizer collaboration facilitated by the Genetic Association Information Network.

The NIH priority that Collins highlighted most often, though, was personalized medicine. Part of the reason for that was timing; Collins's book on genomic medicine, which he wrote after leaving the Genome Institute, came out in early 2010. Media interest in it gave the NIH director a platform to sell the virtues of personalized medicine from a very personal perspective. In *The Language of Life: DNA and the Revolution in Personalized Medicine*, Collins described how he went on his own genomic journey, sending off his DNA to several genotyping companies like 23andMe. The benefits of personalized medicine, though, were not just for famous scientists. Collins assumed control of the NIH just as the United States embarked on a nationwide conversation about health-care reform, leading ultimately to President Obama's Patient Protection and Affordable Care Act. Personalized medicine, Collins claimed, was essential to that transformational legislation; it had the ability to simultaneously cut health-care costs and improve patient outcomes by replacing the one-pill-fits-all approach with one that delivered the right drug, to the right patient, at the right time.

The NIH, on Collins's watch, was going to play a lead role in making personalized medicine a reality. One way to achieve that was by emphasizing newer programs at the agency that catered to the centrality of medical genetics and pharmacogenomics, such as initiatives aimed at identifying genes associated with rare diseases. Another way to do that was by pointing older programs in the direction of personalized medicine. The NIH's large, longitudinal cohorts were a perfect fit for that move, he judged. Existing projects like Framingham and the National Children's Study, Collins said, were natural resources for personalized medicine research. The federal government invested heavily in a national highway system starting in the 1950s, and it fundamentally altered transportation, commerce, and family mobility across the United States. Health care in America was overdue for its own fundamental transformation, and the road to that revolution required federal investment in a "national highway system for personalized medicine."

But the National Children's Study, remember, was not designed to be a resource for personalized medicine. Far from it. After Collins and Duane Alexander abandoned the collaborative American Family Study and parted ways, Collins's American Gene Environment Study was the nationwide cohort explicitly crafted for that purpose. All the design decisions surrounding the National Children's Study, on the other hand, were made oriented around identifying features of the environment that impacted public health and illness across the population of America's youth: the focus on physical, chemical, biological, and psychosocial environmental causes of health and illness; the hypotheses about things like asthma, autism, and diabetes; the plan for collecting the data in the children's homes, schools, and neighborhoods, and the frequency of those visits; the sample size of participating youth; the national probability sample that allowed for making inferences about children across the nation from the representative group included in the study. Those pieces all fit intricately together. No one, not even the director of the NIH, could just snap his fingers and make something like the National Children's Study a personalized medicine resource. If genomics was to go hammering there, it would require significant work turning it into a nail, and so it began.

The first change that came was the subtlest, but it laid the foundation for everything that followed. Alan Guttmacher, Collins's former deputy at Genome and the new head of the Child Health Institute, told the National Children's Study Federal Advisory Committee in 2010 that it was now best to think of the project as a "data-gathering platform," not as a study "designed to answer a specific set of hypotheses." This "hypothesis-free" formulation for doing science became increasingly popular among geneticists as they relied more and more on genome-wide association studies. Such studies involved examining the entire human genome for associations with a trait of interest, without any prior commitment to whether the single nucleotide polymorphisms might be located on chromosome 14 or 6 or somewhere else entirely. That made sense in genetics, where the technological advances in genotyping and sequencing allowed one to scan the entire genome, and so the logistical goal, ever since Neil Risch and Kathleen Merikangas proposed the idea, was to get that data from as many people as possible. But it didn't fit so neatly in the environmental health sciences, because you couldn't just genotype or sequence the entire environment, despite what advocates of the "exposome" imagined. This science started with hypotheses because doing so provided the necessary guidance regarding what data to collect, where to collect it, and how to do so; it also dictated how many people to enroll because different hypotheses required different-size samples to accurately test the theories about how the world worked. What looked like a mere terminological switch—from a "hypothesis-driven study" to a "data-gathering platform"—in fact untethered the National Children's Study from both its ideological and its methodological foundation, opening the door to additional alterations.

The next item to be morphed was the data itself. "Formative research projects" were a whole category of efforts within the National Children's Study initially conceived as a tool for trying out new science and technology. The idea, as initially conceived when Alexander led Child Health, was that a site or sites could try out some new way to collect and process placentas or measure mental and motor development in infants; if a tested method worked well and it was affordable, then it could be expanded to the whole study. Throughout 2011, though, those projects took a decidedly genomic turn. New projects came online with names like "Developing the Capacity for Genomic Medicine in the National Children's Study"; sites were shifted toward genetic analyses. "Genetics, Genomics, and Epigenomics" became an entire class of formative research projects. Had the National Children's Study core hypotheses still been central, this shift would have had to have been justified in terms of which hypotheses that data informed. But once the study was reconceived as a hypothesis-free data-gathering platform, genomic data was easy to prioritize because the technological advances of the previous decade made it relatively cheap and easy to gather.

In the summer of 2011, the new study director, Steven Hirschfeld, made another big announcement. He told the advisory committee in July that the target sample size for the National Children's Study was more than doubling from 100,000 to 250,000. If the goal was to study rare diseases, find rare genetic variants, and uncover common genetic variants with very small effects, then the larger the sample, the better. From an environmental health perspective, though, the sudden increase was troublesome. With finite resources, more participating kids meant less data from each child, and with the turn toward genomic data in the study it looked as if it were going to be the more labor-intensive environmental data that would take the hit.

The huge bump in sample size panicked the scientists and advisers affiliated with the National Children's Study. Ellen Silbergeld and Jonas Ellenberg, two prominent epidemiologists on the Federal Advisory Committee with decades of renowned study design experience, wanted to know why that decision was made, especially because it was done without the advisory committee's input. They were told that it was to facilitate studying rare conditions, but Silbergeld reminded the study leadership that the study wasn't originally conceived to study rare diseases; it had been designed to investigate the common health problems that were having the largest impact on the nation's children. It looked as if the study's goalposts were moving. Half a dozen directors of recruitment sites also privately confronted both Hirschfeld and Guttmacher at the next study gathering in August. They were unnerved by what they were seeing, they explained, developments and shifts that put the study at existential risk. It was partly about the decisions themselves—the jettisoning of the hypotheses, the new sample size target—but it was also about the way those decisions were being made. Under Alexander and Peter Scheidt, the National Children's Study incorporated input from hundreds of disciplinarily diverse experts through the various working groups, advisory committees, and recruitment site personnel. All that changed with the new leadership. The working groups no longer met. The study leadership at Child Health eventually stopped attending the Interagency Coordinating Committee meetings. And the Federal Advisory Committee meetings, rather than serving as opportunities to seek expert advice, became occasions when the Child Health leadership told the advisers what was going to happen next. Study affiliates took to secretly calling Hirschfeld, who had an unusual habit of wearing his Public Health Service uniform at all gatherings, the "Captain," as in "Can you believe this latest directive from the Captain?" The site directors now wrote to the new leadership, imploring them to recommit to the original, core mission of the National Children's Study.

There was no reply. Instead, rumors began to circulate in early 2012 that the study would no longer utilize a national probability sample. As the community of human geneticists moved toward seeking larger and larger research cohorts, they simultaneously sought to avoid the logistical challenges that came with trying to enroll a national probability sample; rather, the goal was to have a research cohort that was as large as possible. Once again, while that worked fine for genetic studies, it severely compromised the ability to investigate social and behavioral contributions to health and make reliable generalizations to the population as a whole.

Now a dispute that had been playing out in private was thrust into the public sphere. Silbergeld publicly resigned the first week of March, telling the new leadership that they had abrogated the intended goals of the children's study. She explained to the press how she and the advisory committee had been sidelined by the new leadership. Ten days later, Ellenberg resigned. He said that the vision of the National Children's Study that took shape after 2009 no longer matched what the National Academies endorsed in 2008. He called for the independent body to undertake a second review. "If we're not there to give advice on this magnitude of change," Ellenberg told a reporter, "I don't know why we're there."

There was more of an exodus from the study. Sarah Keim and her husband, Mark Klebanoff, who worked at Child Health as well and shaped the study from its very origin, left the NIH in 2010. Alan Fleischman, five years after brainstorming how to save the study with

Philip Landrigan and Edward Clark, also left in 2010. Landrigan and Clark, for their part, were informed in February 2012 alongside all the other Vanguard Center directors that they were being phased out in exchange for a plan to hand the recruiting over to a private contractor, forcing them to go back to their institutions and fire or relocate the staff they'd brought on to run the study. The energy of all the federal and academic scientists who'd invested years in making the National Children's Study a reality was being sapped. The general sentiment was that they were leaving behind something irredeemably broken.

Collins appeared before the Senate Appropriations Committee just a few weeks after news of the sampling decision leaked, answering questions in person and responding to written queries that supplemented the discussion about plans for the next fiscal year at the NIH. "Dr. Collins," Senator Richard Shelby, the Republican from Alabama, told the director, "I am hearing serious concerns from the research community regarding proposed changes to the National Children's Study." Shelby and his colleagues voiced worries about the move away from a hypothesis-informed design, the extent to which the results of the research would generalize to children in underrepresented rural areas of the country, the lack of transparency regarding how the decisions were being made. Collins justified the adjustments as cost-savings efforts. But the National Children's Study scientists and advisers weren't complaining about the cost savings; they were accustomed to conversations about cutting costs. They were complaining about the way that the cost savings were sought—by morphing the study into something that it was never intended to be.

The public scrutiny that kicked off in early 2012 only grew in the ensuing months. As it did, attention shifted to the rising price tag of the study. The costs of the study were relatively small in the early years (about $10 million annually), when the focus was on designing the study rather than running it. The expenses increased in 2007 when Congress funded the study and plans for recruitment began. By 2009, when Alexander and Scheidt were removed, the annual expenses had increased to more than $170 million and were expected to stay at that level for several years as the costly recruitment phase of the study unfolded before they dipped down again. When the new leadership at the NIH inherited the project, the study was essentially put on pause while it was being overhauled. And yet, during the pause, they never dialed back the annual expenses; in fact, they increased them to nearly $200 million a year, financing items like the purchase of expensive genetic sequencing and genotyping machines. The new leadership in three years had spent more on the National Children's Study than Alexander had spent on it in a decade. At that rate, the study was scheduled to exceed $1 billion in total taxpayer dollars in 2013. With the study effectively in reset mode, the enormous price tag raised concerns about a federal boondoggle. Congress, in March 2013, took Ellenberg's advice and asked the National Academies to return and figure out what was going on.

After the National Academies convened its new panel, the group planned a process similar to the one followed five years earlier—requesting documentation from the NIH about the current design, reviewing those materials, communicating with the leadership when they had questions. According to Constance Citro, on the staff of the National Academies both times, the difference between the first and the second rounds was "night and day." In 2007–8, the institute provided their reviewers with more than seven hundred pages of detailed documents; they were responsive to inquiries, adding additional materials when they were requested; they welcomed the input from the panel. The second time, the group received just over fifty pages of documentation, pages that described the project in vague, abstract terms. When the panel pressed the NIH for more information about how the study was actually going to proceed, they couldn't get what they needed or even a straight answer about why they weren't getting it. The communication was so poor that the panel couldn't tell if the leadership at the NIH were intentionally stonewalling or if they simply didn't understand what the independent body required in order to do its job.

Some members of the National Children's Study community who lived through these tumultuous years were convinced Collins and Guttmacher were trying to intentionally sabotage the project. In fact, the geneticists were trying to save it in the only way they knew how. All the major adjustments that had been made to the study between 2010 and 2012 created a Frankenstein's monster, where the environmentally focused public health project called for in the Children's Health Act was being pieced together with something that attempted to look more and more like Collins's American Gene Environment Study. Faced with the public insurrection that was unfolding, the new leadership at NIH gave up on some of the switches, tried to find a middle way for others, and introduced entirely new ideas elsewhere. The sample size, for example, went back down to 100,000. The original hypotheses that took years to develop had been discarded, but suddenly a handful of "exemplary hypotheses" were introduced to "inform, rather than define, the study's design," language that Collins used in AGES. The population of participating children, which they'd switched from a national probability sample to a convenience sample, was instead going to be part probability/part convenience.

The National Academies delivered its assessment to the NIH and members of Congress in June 2014. They opened by affirming the importance of the study called for in the Children's Health Act, saying the effort had "the potential to add immeasurably to scientific knowledge about the impact of environmental exposures, broadly defined, on children's health and development in the United States." The rest of the report was one long indictment. The main theme was the panel's inability to judge the study design decisions because of how little information they were provided. On the planned use of exemplar hypotheses: "the panel finds that the exemplar hypotheses proposed for the [National Children's Study] are not sufficiently well developed to guide sample design and data collection in the early waves, nor is there a plan to identify lines of inquiry that could lead to important exemplar hypotheses to guide data collection in later waves." On the sample of children: "the documents that [the National Children's Study] made available to the panel did not provide sufficient details for an evaluation of whether the proposed sample would meet the minimal standards of a scientifically based sample design required for large national data collections." On the data planned for collection: "since the panel did not receive information on specific study protocols, data collection methods, or study instruments, its review could not address the scientific merit or quality of these aspects of the [National Children's Study] data collection." On the frequency with which they planned to collect the data: "the panel did not receive an adequate explanation of the scientific basis for the specific proposed schedule of visits."

In a highly unusual move, the panel also included an appendix documenting its difficulty communicating with NIH during the course of its review, as well as an entire chapter on problems with the leadership and oversight of the National Children's Study. The panel wasn't charged with reviewing the study leadership and oversight, so saying anything about those elements in the report was risky because it could have been interpreted as unauthorized mission creep. But they felt obligated to do so because of the dysfunction they discovered. "The study leadership were trying to do something that they really weren't qualified to do," Citro surmised, "and they didn't recognize that they weren't qualified to do it."

Just what to do about the National Children's Study was unclear. The panel endorsed the general vision and importance of the study called for in the Children's Health Act, but it was hard to see how the NIH personnel in 2014 could deliver that. Collins, upon receiving the report, put the study on hold and then, in July, convened his own Advisory Committee to the Director charged with judging whether the National Children's Study was "feasible, as currently outlined, especially in light of increasing and significant budget constraints."

The experts that Collins gathered took their task seriously. They were aware that a great many scientists and scientific communities—people and groups they knew personally—had invested years in the National Children's Study. They tried interpreting "feasible" in different ways to see if alternate interpretations led to different judgments. They took turns arguing for and against the study's viability. Ultimately, though, the result was a foregone conclusion. Like the National Academies panel before them, this new group admitted that they had difficulty articulating what the plan for the study was in 2014, let alone whether that plan could be achieved. The group issued its verdict in December: the nationwide study of America's children called for in the Children's Health Act of 2000 "as currently outlined, is not feasible."

Eleven years after Collins sat down on the couch in Alexander's office, Collins terminated the project he'd asked to piggyback. The

NIH director said he was "disappointed that this study failed to achieve its goals." At the same time, he didn't want the nation or the scientific community to fear that he'd lost faith in large, longitudinal cohort research more generally. On the contrary, "I am optimistic that other approaches will provide answers to these important research questions."

7

DNA's "Dirty Little Secret"

WHEN MARYANNE DICANTO, of Doylestown, Pennsylvania, felt a lump in her breast, she scheduled an appointment to have it examined. At the clinic, the physician told her that the mammogram was inconclusive. There was a growth, but her physician couldn't tell whether it was a benign fibroid or a malignant tumor and wanted to perform an MRI to get a better image. The problem was that DiCanto's insurance wouldn't cover the more expensive procedure. DiCanto refused to accept that decision. So she and her husband, Scott Primiano, walked out to their car, got on the phone, and just kept at it until a representative at her health insurance provider heard her out and reversed the decision. DiCanto and Primiano opened the car doors and walked back into the physician's office, where DiCanto slid into the noisy magnetic rings of the MRI machine.

The couple were vacationing on Chincoteague Island, Virginia, a week later when MaryAnne received the news that the MRI confirmed it was no benign fibroid. The diagnosis was both shocking and somewhat expected. DiCanto's mother, Mary, had had breast cancer for thirty years. MaryAnne knew that she was at increased risk and was hypervigilant about monitoring for any sign of the disease. Still, it's one thing to live with that possibility and another to get the phone call.

DiCanto and Primiano left Virginia and met again with her physician to plan the medical response. They proceeded with an aggressive course of treatment—a double mastectomy, radiation therapy, general chemotherapy, and a prescription for tamoxifen, which reduced the risk of recurrence. The chemotherapy made DiCanto so hot that she would stand outside in the slushy Pennsylvania snow to cool off. Still, she never wavered. She admired her mother's determination in the face of her own experience with cancer, and like her mother, DiCanto was committed to not being victimized by the disease. She followed a basic game plan: do what her physician asked, follow the treatment schedule, eat healthy foods, and exercise, and things were going to work out. That was in 2003, and for a decade it worked.

DiCanto and Primiano were constantly aware that the disease could come back. Every fever, every cough, every wheeze, was viewed through the lens of its return. "Cancer was a family member," Primiano recalled. In 2013, his wife felt a nagging pain in her side. Living on Long Island by then, DiCanto and Primiano visited another doctor, in New York. An X-ray, an MRI, and then a bone biopsy confirmed their worst fear: the cancer was back, and it had already metastasized, invading DiCanto's ribs, lymph nodes, and liver.

Hearing "metastatic cancer" is particularly terrifying, often interpreted as a death sentence. DiCanto, however, was once again comforted by her mother's example. She was also diagnosed metastatic in the 1980s but went on to live for fourteen years. If her mother could survive for that long three decades ago, DiCanto figured, she could do at least as well with the medical developments that had emerged since then. She immersed herself in the latest science—reading research articles by oncologists, joining online groups like METAvivor and the Tutu Project, where patients and their families shared information about new clinical trials, and investigating new approaches to cancer care that treatment centers advertised.

It gave DiCanto and her husband real hope. A lot had changed in the world of breast cancer treatment since her first experience a decade earlier. The excitement now was all about genomic testing. Unlike the previous treatments that she received, which attended to the symptoms of cancer, commercials advertising genetic testing claimed that the game-changing technology got at the underlying causes of the disease. And now the debilitating side effects of chemotherapies could be mitigated because a patient's own DNA indicated which drugs worked for her and which didn't. DiCanto was determined to take advantage of this new genomics, which clinicians, biomedical researchers, and private companies were calling "precision medicine."

It was not surprising that DiCanto found widespread excitement about it. What geneticists had been referring to as "personalized medicine" when she was diagnosed the first time was routinely packaged as "precision medicine" now. The story of how personalized medicine became precision medicine is a tale about how well-intentioned efforts to clear up misunderstandings regarding genomics managed only to compound the confusion, with harsh consequences for patients desperately seeking answers.

. . .

Personalized medicine, judged purely through the lens of a marketing campaign, was an astonishing success. Geneticists like Gualberto Ruaño, recall, pushed in the late 1990s to incorporate genetics into the drug developing, drug testing, and drug prescribing processes. That took some convincing, though, because the drug industry had traditionally embraced blockbuster drugs that were intended to work for everyone. Only when adverse drug reactions became common and publicized did the pharmaceutical world come around to the opportunities presented by pharmacogenomics: the "right drug, right patient, right time" alternative. Genomic profiles could save patients from harmful side effects, guide doctors when it came time to write prescriptions, and refine the research and development pipeline for drug companies. "Personalized medicine," though it started out simply as the slogan of Ruaño's Genaissance, became the name of that revolution.

The branding was so successful that it quickly became adopted by practitioners of genomic medicine more generally, even when the clinical applications had nothing to do with drugs. Preimplantation genetic diagnosis of human embryos, family screening to check for a risky gene that increased the chances of developing cancer, genetic testing by parents to see if they carried a harmful stretch of DNA that might be passed on to a child—all of these things and others gathered under the expanding umbrella of personalized medicine. Practitioners of other "-omics" medicines that tracked molecules closely related to genes—proteomics, transcriptomics, metabolomics—also hopped on board. Promising to treat patients as individuals became the new craze that then generated its own momentum. The pharmacogenomic marketing language, conveniently, had to change very little to accommodate the expansion. "Right drug, right patient, right time" just became "right treatment, right patient, right time," and the contrast with one-pill-fits-all pharmacology morphed ever so slightly into a contrast with one-size-fits-all medicine. Personalized medicine journals were created, and academic citations of the term increased from a couple to hundreds per year. Universities set up personalized medicine programs to train the next generation of physicians in the new approach to health care. The Washington, D.C.–based Personalized Medicine Coalition formed in 2003 with the goal of educating policy makers about the promises of personalized medicine, uniting stakeholders invested in the practice, and advocating for a federal regulatory environment that encouraged growth. Champions of the medical innovation heralded a new era of biomedical game changers and miracle cures.

All that success brought scrutiny. Critics, many of them practicing physicians, clinical researchers, and even prominent pharmacogenomicists, raised three concerns. Personalized medicine was judged to be *misleading, offensive,* and *overhyped.* It was misleading because personalized medicine was not personal. Advocates of personalized medicine characterized the science as if it produced "individualized drug therapies" and an "individual, genetics-based approach to medicine." Reporters who interviewed the scientists wrote articles with titles like "Personal Pills" and "Just for You." The picture that emerged was one of personalized medicine turning doctors into boutique tailors who fabricated bespoke pharmaceuticals specially crafted for each unique individual. But that wasn't at all how pharmacogenomics in fact worked. Pharmacogenomics grouped people based on shared genomic profiles that indicated how their bodies metabolized different drugs—those who had the epidermal growth factor receptor mutation from those who did not, or those who were primaquine-sensitive from those who were not. This wasn't individualized, bespoke tailoring. It was the garment equivalent of being able to pick a shirt off the rack that was either small or extra large.

Some practicing physicians also took issue with the suggestion that geneticists were the only health-care providers in the business of avoiding one-size-fits-all medicine. In fact, doctors had known for millennia that what was good for one patient might not work for another. The Hippocratic Corpus, that collection of early medical writings from the age of Socrates, is filled with advice regarding the need to keep individual differences between patients in mind when diagnosing and prescribing. Two people with fevers, for example, might need very different treatments: "Give different ones to different patients, for the sweet ones do not benefit everyone, nor do the astringent ones, nor are all patients able to drink the same things." Physicians can find advice regarding the natural variation in shoulder sockets, which are crucial to mind when it comes time to reset a dislocated joint, as well as individual differences in the course of pneumonia. There is a great deal in the Hippocratic Corpus that you would not want your health-care provider to prescribe for you today. Much of the medical practice was concerned with diagnosing perceived imbalances in black bile, yellow bile, blood, and phlegm (the "four humors") and then treating those imbalances with things like bloodletting and vomiting. The notion, though, that medicine must be attuned to the particular features of a patient remained a core element of health care for the next twenty-five hundred years.

When a patient needs a bone marrow transplant, one of the factors that the health-care team will take into consideration is the HLA match between donor and recipient. "HLA" stands for human leukocyte antigen complex, and it's a reference to proteins on the cells of a person's body that the immune system uses to distinguish biological friend from foe. To ensure that a patient's immune system will accept the new tissue, physicians ensure that the HLA protein profile of the donor is close enough to that of the recipient that the tissue will not be rejected. A similar but more familiar example is the incorporation of blood type into transfusions. The Viennese researcher Karl Landsteiner first identified different blood groups in humans in 1901, for which he later won the Nobel Prize. Like HLA matching, pairing the blood type of a donor to a recipient is essential during blood transfusions; if you are type O but given a type AB donation, severe hemolytic anemia ensues as your immune system attacks the foreign red blood cells, which can lead to kidney failure and even death. Careful attention to the particular features of patients also goes into the administration of general anesthesia. When a patient undergoes surgery, an anesthesiologist will collect personal information about smoking and alcohol use, prescription drugs and supplements, as well as known allergies, before delivering the anesthetic; then, during surgery, the anesthesiologist actively manages the anesthetic level in response to the patient's heart rate, blood pressure, breathing, and oxygen levels. A physician who gives two diabetic patients unique treatment plans based on differences in their financial resources, social support, insurance coverage, and responsiveness to past recommendations is also avoiding one-size-fits-all medicine.

These are just a few examples of how health-care providers orient their practices around what makes one patient different from the next. But they don't call them "personalized medicine." They call them "medicine" because that's what doctors do. So it was insulting to physicians everywhere when champions of personalized medicine came along and suggested that health-care providers were stuck in an old-fashioned mindset unless they were tracking genes. In fact, it was the pharmaceutical industry that had become committed to that mindset with its blockbuster drug model. Pharmacogenomics represented something genuinely innovative within the drug industry, but it was not so revolutionary when it migrated out of that domain and into medicine more generally, where attending to patient differences was the norm. It was only when and where health care became increasingly reliant on pharmaceutical responses to health concerns that the one-size-fits-all mentality appeared to be so widespread in medicine.

As if those two problems weren't enough, some of the leading voices in the field of pharmacogenomics stepped forward and expressed concerns about how advocates of personalized medicine were overselling expectations regarding what the science could reasonably deliver. Daniel Nebert, one of the pioneers of research on the genetics of cytochrome P450 function—the enzyme that ushered in the era of mainstream pharmacogenomics—warned against claims that a pharmacogenomics revolution was just over the horizon, because often dozens or even hundreds of genes were implicated in drug responses. The genomic complexity alone would take many years to unravel, he explained, and that didn't include all the environmental factors that impacted the metabolism of pharmaceuticals. Arno Motulsky urged similar restraint. As the language of personalized medicine was becoming common parlance in medical genetics, the discipline he helped create, Motulsky worried, "What we know about the genome today is not enough for all the miracles many expect from this field."

This consternation surrounding the concept of personalized medicine was only growing when some of the most prominent figures in human genetics at the time—stars from the Human Genome Project, scientists who helped develop gene-targeted drug therapies, pharmaceutical executives who led divisions of genomics—gathered in 2011 to make the case for the central role of genomics in the future of health care. They congregated in Washington, D.C., at the request of Francis Collins, who asked the National Academies of Sciences, Engineering, and Medicine, the same organization (albeit constituted by different committee members) that reviewed the National Children's Study in 2008 and then raised red flags about it in 2014, to convene a committee focused on reimagining health and illness at the level of molecular biology. The report that the group issued reads like a textbook on personalized medicine. It foresaw a "genomic transformation of medicine." It explained the basics of pharmacogenomics. It proposed a massive study of DNA from a million Americans to link up genetic differences with differences in health outcomes. But when it came time to give their final report a title, they opted against *Toward Personalized Medicine.* They called it, instead, *Toward Precision Medicine.*

The choice of "precision medicine" was part of an intentional effort to move away from equating genetics with actual, individualized care and instead to highlight a different anticipated strength of genomic medicine. The ancient Greek philosopher Plato, a contemporary of Hippocrates's, offers the resources for understanding that feature. In *Phaedrus,* Plato tells the story of Socrates guiding his young pupil Phaedrus in how to think about the world. One principle for good reasoning, Socrates explains, involves "that of dividing things by classes, where the natural joints are, and not trying to break any part, after the manner of a bad carver." The focus of *Phaedrus* is love, and so the dialogue between Socrates and his student revolves around things like the similarities between romantic passion and madness, and the dangers of mixing friendship with sex. Plato's analogy of carving nature by its joints, though, left a lasting impression far beyond philosophical reflections on love. It's a reference to the difference between an experienced and an unexperienced butcher. An experienced butcher knows exactly where the natural joints of a carcass reside and guides the knife along those lines such that the cuts of meat separate cleanly and effortlessly. A bad carver fights the carcass, forcing unnatural divisions on the body by way of a messy struggle. Good science too, the modern idea goes, carves nature at its joints. It finds the natural division of things rather than imposing artificial constructs.

Most of medicine, the authors of *Toward Precision Medicine* claimed, approached the study and treatment of health with a vision of the human body that was divided up by Plato's bad carver. Cancers were crudely distinguished based on which organ or tissue they impacted—blood, bone, breast. Illnesses were superficially categorized based on their signs or symptoms—fever, infection, hypertension. An experienced butcher, however, had opened up shop, and this butcher divided up the human body differently, along the natural joints of molecules and genes. What was previously thought of as one organ-level cancer such as lung cancer could be chopped up as a dozen distinct cancers with unique genetic profiles and unique molecular pathophysiologies. What's more, one genetic mutation might be associated with molecular processes that cut across many different organs. Treatment of symptoms, it was argued, was being replaced with precision medicine's intervention on causes.

The authors admitted that they were using the term "precision medicine" just as others had been using "personalized medicine" for more than a decade before them. Both were about tailoring medical treatments. But personalized medicine had been causing confusion, and they wanted to avoid that. Whereas personalized medicine misleadingly suggested that each patient would get her or his own unique treatment, precision medicine was about replacing diagnostic ambiguity and therapeutic uncertainty with genomic accuracy and molecular clarity.

Because *Toward Precision Medicine* came with the prestigious imprimatur of the National Academies, scientists and nonscientists alike took notice. Health journalists reported on the terminological shift that was afoot. Now it was precision medicine journals appearing, and citations to *that* term exploding. Now universities began adding precision medicine programs, centers, and institutes. Conferences aimed at hospital administrators, pharmaceutical executives, and business strategists emerged with a focus on implementing precision medicine in health care. Hospitals began advertising "precision oncology" as their innovative new approach to cancer care. The switch even crept out beyond the health-care landscape: proponents of "personalized education" began repackaging their educational genomics as "precision education." This was the world that MaryAnne DiCanto found herself in when her breast cancer returned in 2013.

. . .

DiCanto threw her body and soul into combating metastatic cancer. She followed the latest news regarding precision medicine therapies. She asked her doctor about enrolling in clinical trials of heralded drugs. Just as my father was prescribed erlotinib after the results of his biopsies, DiCanto underwent a series of excruciating lung and liver biopsies, where tiny clumps of tissue were cut from her body and used for DNA analysis, designed to identify which of her genetic markers were good targets for either FDA-approved or still-experimental drugs.

What she and her husband got in return was a great deal of frustration and heartache. In one case of precision medicine, a genetic test paired DiCanto with everolimus (brand name Afinitor), a drug designed to interfere with the growth and spread of cancerous cells. But after taking the medication for just a brief period, she became terribly weak. She and Primiano rushed to the emergency room, where they learned that her red blood cell count had plummeted to dangerously low levels. DiCanto was experiencing anemia, a well-known potential side effect of everolimus. The whole purpose of precision

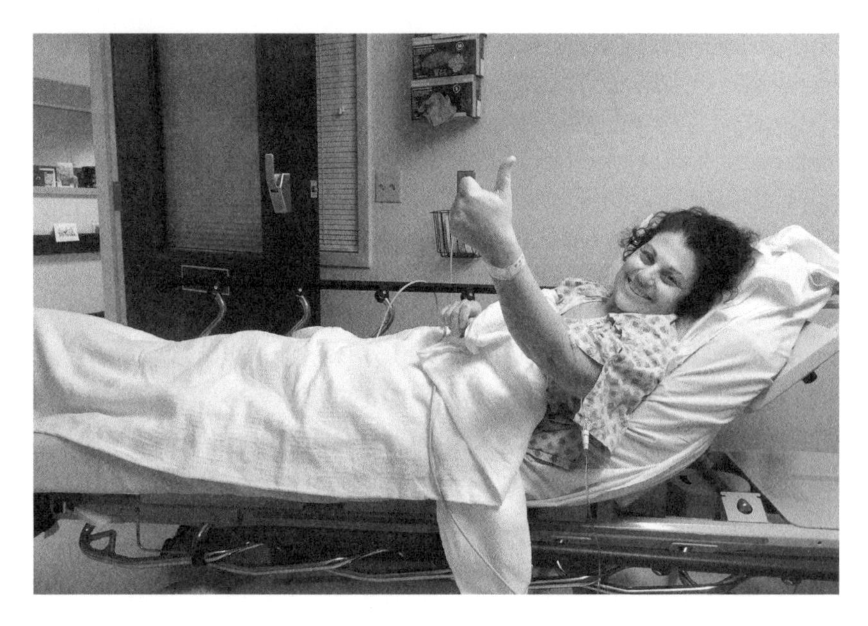

MaryAnne DiCanto, in 2015, preparing for biopsy. Photograph by Scott Primiano

medicine, as DiCanto and Primiano understood it, was to avoid catastrophic side effects by finding the precise gene-drug matches for each patient. And yet here was DiCanto on a drug that her genomic test indicated was a good pairing, and it nearly killed her.

Their frustrations were compounded when the results of subsequent genetic testing didn't align with the results of previous tests of her DNA. Different tests by different companies at different times produced different results about different genomic markers. Some showed lots of potential drug targets, others few, and they didn't always line up. That left DiCanto and Primiano angry and confused. According to Primiano, "Things weren't getting more precise; they were getting less precise." He felt as if his wife suffered greatly, only to pull back the curtain on the wizard, where they saw a reality that was hardly a revolutionary new approach to cancer care. "You think it's going to be precise, like a laser versus a shotgun," he told the health reporter Liz Szabo, who investigated DiCanto's story for Kaiser Health News; "but it's still a shotgun."

DiCanto and Primiano were disappointed, but what choice did they have? DiCanto was desperate. "You'd stand on your head for two hours each day if somebody told you it might combat cancer," Primiano said. Cancer patients, particularly those with metastatic cancer, are always looking for some sign of hope. DiCanto and Primiano didn't talk about her cancer being terminal. They talked about what they were going to do to keep fighting—the next test, the next treatment, the next clinical trial. DiCanto never stopped. That same dogged determination that had her phoning up insurance representatives and demanding an MRI back in 2003 was still there fourteen years later when she tried her last round of genomic testing in 2017. DiCanto, her body shutting down, went in for yet another treatment indicated by a genomic test on a Thursday and died the following Saturday. She was fifty-nine.

Primiano bears no ill will against the science of genomic oncology per se. It's the packaging that infuriates him—the cancer centers that only show faces of smiling survivors, the promises of a precision medicine that uses DNA to find the perfect gene-drug matches. How could something called "precision medicine" be anything other than precise? If the commercials, genomic testing companies, and oncologists had simply said, "There's this test, and it might do some good with regard to ruling out certain treatments that are out there, but you also have to keep in mind that it isn't perfect, and the science is evolving, so the test results very well might evolve too," then he and DiCanto could have waded into that process with more realistic expectations. Instead, he said, it created false hope. To this day, when cancer center commercials come on television or radio advertising precision medicine testing and treatments, Primiano tells anyone who will listen, "That. Is. Bullshit." Still, he knows most people don't want to hear that. They want to hear about the inspirational survivor stories and the miraculous cures and the pervasive language of precision medicine that reinforces the aura of success.

. . .

The switch from personalized medicine to precision medicine was supposed to clear up confusion regarding what genetics actually brought to health care. The new terminology, however, suffered from the very same faults as its predecessor: it too was misleading, offensive, and overhyped. DiCanto's experience exemplified these problems, but her story was by no means an anomaly. In recent studies that focused on patients with advanced cancers like DiCanto's, it turned out that only 5–7 percent of the patients benefited from drug matching following genomic testing, and the benefit was a few months added to the end of life, much like my father's experience. DiCanto's frustration with multiple genetic tests producing conflicting results was also not unusual in the world of oncology. Researchers at the University of Washington enrolled patients diagnosed with cancer in a study to see what would happen when samples from each patient were sent to two different genomic testing companies that specialized in matching cancer patients with gene-targeted treatments. Of the forty-five genomic markers from the patients identified by the tests, only ten were picked up by both tests. And of the thirty-six drugs that the tests suggested for the patients, only nine were recommended for the same patient by both companies. In some patients, there was no overlap at all between which drugs the two companies recommended.

This lack of precision is not confined to the end of life. It can also create confusing chaos at the beginning. In 2011, Esmé Savoie showed signs of respiratory problems and low muscle tone shortly after she was born, leading to a week in the neonatal intensive care unit. In the ensuing months, she experienced violent vomiting and aspiration that resulted in cardiorespiratory arrest, as well as debilitating seizures. Relief finally seemed to come for Esmé and her family when she was about to turn two and genetic testing revealed a rare mutation in *PCDH19*, a gene involved in cellular communication in the brain and known to be associated with childhood epilepsy when dysfunctional. Esmé's family took solace in this genetic explanation, and they immersed themselves in the *PCDH19* community. Her mother, Hillary, who documented it all on a blog, joined an online support group for parents and children with *PCDH19* epilepsy. She attended academic meetings where scientists presented the results of their research on the rare condition. She even started a foundation and took Esmé to visit a zebrafish lab that received a grant from the organization to begin studying *PCDH19* biology in the model organism.

As Esmé grew, however, differences between her and the other kids with *PCDH19* epilepsy became more and more apparent. The other

Hillary Savoie with her daughter, Esmé, in 2016. Michael P. Farrell/*Times Union*

children were walking and running while Esmé still couldn't sit up. Convinced by Esmé's physicians and the *PCDH19* researchers that it would be worth undertaking a new round of genetic testing, her parents agreed in 2015. This second round of testing both undid the results of the first test and sent Esmé's family looking in an entirely new direction. Esmé's *PCDH19* gene was no longer deemed pathogenic. The new culprit was a different gene—*SCN8A,* another segment of DNA implicated in childhood epilepsy that is even rarer than *PCDH19* epilepsy. So her mother shifted the focus of the foundation from *PCDH19* to *SCN8A* research and began interacting with online *SCN8A* groups. The new diagnostic home felt like a better fit, but Hillary balanced that comfort against the newfound uncertainty surrounding the stability of these genetic tests.

Wisely, it turned out. In 2016, the same company that provided Esmé's 2015 genetic test result updated her genetic status again. Now it wasn't sure about the pathogenicity of her *SCN8A* gene, but it did find two new genes that looked problematic—*TBL1XR1* and *MAP3K7.* At this point, the explanation of Esmé's rare condition remains a clinical mystery; her particular set of symptoms doesn't perfectly match any of the genes implicated.

Esmé's condition may be unique, but, like DiCanto, her experience with genetic testing was not. In the world of rare, pediatric epilepsies, alterations in the classification of genetic variants' pathogenicity, which can then change clinical diagnoses, is not unusual at all. A 2019 study from researchers at the University of Texas Southwestern reviewed three hundred cases of epilepsy over a five-year period. In that time, nearly a third of the children had their diagnoses changed as a result of updated genetic testing. Esmé's mother described the experience as coming to discover the "dirty little secret" of DNA: genetic tests "come with asterisks attached." Those asterisks take their toll. Some people questioned whether the family was just making the whole thing up. How, they demanded, could genetic conditions come and go? The clinical uncertainty exacerbated an already stressful environment. Caring for a child with Esmé's needs is an exhausting and anxiety-inducing exercise in unconditional love. What the next week or day or even moment holds is often unknown; genetic uncertainty extends that to the very diagnosis and prognosis, which makes planning for the future extremely difficult.

The stories of MaryAnne DiCanto and Esmé Savoie remind us that genomic medicine is constantly evolving. That, in itself, is not a flaw. It means the science is moving forward and incorporating new information, as it should. As one of the UT Southwestern researchers surmised, "It's a testament to how quickly genetics is evolving and how much we are learning with new laboratory techniques and with more people getting their genes tested." All health sciences proceed in the same way. The flaw is in suggesting that the outcome of this ever-changing process is precision. What seems clear one day can become ambiguous the next. That's not going to stop anytime soon; it's built into the very nature of how the science of medical genetics progresses. The real danger, of course, is that most patients and families don't realize this until after they've entered the world of genetic testing, lured in part by the promise of precision, and only then start receiving the conflicting results.

Precision medicine, like personalized medicine before it, was also insulting to health-care workers. Precision medicine's revolutionary claim was its innovative ability to get at underlying causes, in contrast to more traditional medicine, which only treated symptoms.

But this assertion was just as unfair to health care as was personalized medicine's suggestion that traditional medicine was stuck in a one-size-fits-all mold. Both oversold their novelty by painting a crude caricature of the contrast. The Hippocratic Corpus is all about treating causes, and much of the subsequent history of medicine is about the series of successful efforts at identifying the underlying causes of various health conditions—germs as the agents of infection, blood clots as the sources of stroke, environmental carcinogens as the triggers of cancer, drinking water contaminated with sewage as the cause of a cholera outbreak. Precision medicine doesn't uniquely and revolutionarily get at underlying causes in a way that traditional medicine does not. It simply privileges particular types of underlying causes—molecules and genes. That's fine for conditions that are truly genetic in origin, but most diseases are influenced by many different factors and, in turn, can be combated with treatment and prevention strategies aimed at many different underlying causes. Gene-targeted drug therapies are one treatment for cancer, for example, and they do indeed take a patient's genomic information into consideration to judge the suitability of a treatment. But for the vast majority of cancers the most effective treatment is still surgically removing a solid tumor before it spreads. Likewise, patients who have a tumor excised are often then treated with radiation therapy; this treatment is about killing any remaining cancerous cells that were left behind. Hormone therapy can be used to starve certain cancers, like breast and prostate, that rely on a body's hormones to thrive. Tumors can also be burned with radiofrequency ablation therapy or frozen dead with cryoablation therapy. As for prevention, when someone quits smoking, their risk of developing cancer goes down dramatically. Removing carcinogenic pesticides from the environment reduces rates of leukemia and lymphoma. Tissues, hormones, cells, genes, habits, environments—there is no single target for cancer treatment and prevention because there is no single, underlying cause of cancer. By equating precision medicine with interventions that target genes, the insulting message is that focusing on genes is the only way to precisely get at underlying causes, that the surgeon and the endocrinologist who aren't sequencing genomes are providing their patients with *imprecise* medicine.

The switch from personalized medicine to precision medicine also did nothing to combat the hype surrounding that science. If anything, it only exacerbated it. All the talk about "game-changing drugs" and "therapeutic breakthroughs" persisted. Now it was bolstered by new talk about clarity and accuracy. Companies offering genetic testing advertised "clear information about companion diagnostic genomic findings and biomarkers and their associated target therapies." Cancer care centers added "precision oncology" to their menu of featured services. All this despite the fact that most oncologists continued to say it was too soon to predict the long-term impact of precision medicine on cancer care.

What made all of this even more confusing for patients and their families entering the world of genetic medicine was that talk of personalized medicine never abated. True, when precision medicine burst onto the scene, many academic researchers and clinicians adopted the new language. But there were also plenty of stakeholders who had been marching under the banner of personalized medicine for more than a decade: publishers with journals bearing the title, universities awarding certificates in personalized medicine, patients discussing their experience with personalized medicine on cancer blogs. In fact, interest in personalized medicine only expanded after precision medicine joined the conversation. Precision medicine journals appeared quickly after 2011, but so too did new journals purporting to be devoted to personalized medicine. Academic citations of personalized medicine didn't decline in response to the new phrase; they increased year after year. University programs offering training in personalized medicine multiplied.

Two general strategies emerged in the clinical and research communities to navigate the conceptual conundrum that they'd created. The first involved stipulating some formal relationship between personalized and precision medicine so as to divide up the turf. Alas, the two terms were introduced as synonyms albeit with different emphases, so any distinction would be arbitrary. Indeed, the result was about as many distinctions as fabrications: personalized medicine was the overarching concept and precision medicine was a subset of it; precision medicine occurred first in the clinical process, and then personalized medicine followed; precision medicine was about a clinician's relationship to genetic information, while personalized medicine was about a patient's relationship to genetic information; personalized and precision medicine tracked different clinical processes; precision and personalized medicine tracked different technologies; personalized and precision medicine tracked different biological processes. None of those formulations really made any more or less sense than the next, which made it challenging for an outsider to follow what the geneticists were saying or how what one clinician said matched, or differed from, what another one said. A patient recently diagnosed with cancer and looking for information about the latest treatment options could read one article where she'd be told personalized medicine is what direct-to-consumer companies like 23andMe and AncestryDNA provide consumers while precision medicine is what doctors provide patients; then she'd enter the clinic where she would be told by her doctor that they would be providing her with personalized medicine services, some of which might also be precision medicine technologies.

The other strategy—and this one was both truer to the actual history and more common—involved simply embracing the two concepts as synonyms. Phrases like "personalized medicine, also called precision or individualized medicine," became common, or "precision medicine, also varyingly known as individualized medicine, personalized medicine, or genomic medicine." This inclusive mindset was an effort to allow both concepts to coexist. That harmonious inclination, however, compounded the confusion. A patient newly diagnosed with cancer who made his way to this conceptual landscape was led to believe that genomic medicine personalized treatment (it did not), and it alone treated patients as individuals (that's false), and it replaced ambiguity and uncertainty with precision and clarity (not true), and it alone got at the causes of disease (also not true). Precision medicine was introduced as a well-intentioned effort to clear up misperceptions surrounding genetics, but the result was the exact opposite. That's a dangerous irony for desperate patients and families making their way to a science surrounded by talk of biomedical breakthroughs and miraculous cures.

. . .

The second week of June 2014 was a particularly momentous one for the National Institutes of Health. The National Academies delivered its scathing review of the National Children's Study. On a more positive note, Collins was invited to meet with President Obama for reasons that had nothing to do with the study of kids. The president was looking toward the final years of his presidency and was interested in announcing a major initiative at the intersection of genomics and health, an extension of his investment in personalized medicine from his time as a senator. Collins and the NIH were invited to develop an idea for what shape that initiative could take.

This was the opportunity that Collins had been seeking since 2003. He had tried to take his proposal for the American Family Study, alongside Duane Alexander, to President George W. Bush in his first term but was turned down. He'd offered, without Alexander, the American Gene Environment Study to the secretary of health and human services in Bush's second term, but Michael Leavitt wouldn't provide the desired amount of money. When Collins ascended to the role of NIH director in 2009, he took over the agency in the midst of the Great Recession, when funds for major new initiatives were limited, leaving him stuck trying to force the National Children's Study peg through a personalized medicine hole. Now, with President Obama in his second term and the economic turbulence behind, the NIH director was encouraged to suggest something bold.

Collins gathered a team of his closest and longest-serving colleagues to take the plan that they'd been conceiving for more than a decade and turn it into something suitable for the current occupant of the Oval Office. They sent to the president a two-part plan to give all Americans access to the genomic medicine revolution—first, a major investment in genetic oncology, which was of particular and personal interest to President Obama, and second, a nationwide research initiative designed to recruit a million or more participants into the largest biomedical research program in the nation's history.

In October, as Collins's advisory committee was assessing whether the National Children's Study was "feasible, as currently outlined," the NIH director traveled to the White House to meet the president. The Oval Office is smaller than one might expect, and the sense of history fills the space with a tangible magnitude. Portraits of George Wash-

The October 3, 2014, meeting in the Oval Office, where a proposal to enroll a million Americans in a nationwide genomics research cohort was embraced by President Obama. Pete Souza / The White House

ington and Abraham Lincoln stared down at the guests who took their seats on couches before President Obama; a bust of Martin Luther King Jr. looked on. An enormous rug encircled the sitting space, rimmed with quotations from past presidents. From Theodore Roosevelt: "The Welfare of Each of Us Is Dependent Fundamentally on the Welfare of All of Us."

The room was particularly crowded that day. In addition to NIH representatives, leadership from Health and Human Services, the Food and Drug Administration, the Office of Management and Budget, and the Office of Science and Technology Policy were all in attendance, reflecting the sprawling scope of the initiative under discussion. President Obama, to Collins's delight ("utterly awesome," as he described it), approved of the NIH plan and instructed his staff to make it a budgetary priority for the coming fiscal year. That would permit the initiative to begin taking shape while the president was still in office and then serve as a legacy after he was gone. "We have a tiger by the tail!" Collins reported back to NIH after he'd heard the news.

On January 20, 2015, Collins and his team, which had grown to two dozen after the president green-lighted their plan, gathered at Collins's home in the tony Chevy Chase suburbs of northwest Washington, D.C. They squeezed onto his couch, squatted upon cushions on the floor, and stood along the walls, some nervously sipping red and white wine, others bottles of pale ale. The group had learned in November that the president would announce their plan to the nation in his next State of the Union address, scheduled for that night. Exactly what the president would say, though, was a mystery. Just a passing reference to the need for more federal biomedical research, which only they would know was a nod to their nationwide plan? Or might he mention it by name? If by name, what would that name even be? By 2015, genomic medicine was in the midst of its terminological metamorphosis from personalized medicine to precision medicine, and so the group didn't know what the president's staff settled upon in terms of giving their plan a title.

Thirty minutes into his address, the president folded his hands and pivoted from talking about federal support for business to federal support for health care. "I want the country that eliminated polio and mapped the human genome to lead a new era of medicine—one that delivers the right treatment, at the right time." Collins pointed to the television. It was happening. "In some patients with cystic fibrosis, this approach has reversed a disease once thought unstoppable. So tonight I'm launching a new Precision Medicine Initiative to bring us closer to curing diseases like cancer and diabetes, and to give all of us access to the personalized information we need to keep ourselves and our families healthier. We can do this." Precision medicine, personalized information, the right treatment at the right time, a new era of medicine, curing disease—it was all there. Collins's living room erupted in applause and cheers.

They weren't the only ones. Political polarization only grew during the Obama administration, which was reflected by the fact that most of the president's proposals that evening received hearty approval from the Democrats and stern silence from the Republicans sitting in the audience. The Precision Medicine Initiative, on the other hand, drew a standing ovation from representatives and senators on both sides of the aisle, confirming just how popular genomics had become

Celebration at Francis Collins's home as President Obama announces the Precision Medicine Initiative in his 2015 State of the Union address.
Photograph by Eric Green

across the political spectrum. Now it was time to make the nationwide project a reality.

Between 2015 and 2018, at the same time that Hillary Savoie was learning about the asterisks of genomic testing and Scott Primiano was coming to grips with the fact that more genomic testing meant less precision, the NIH prepared to roll out across the United States what was initially dubbed the "Precision Medicine Initiative Cohort Program." Historical precedent suggested such a project would be housed in one of the NIH's various institutes—Framingham in the National Heart Institute, the Collaborative Perinatal Project in the National Institute of Neurological Diseases and Blindness, the National Children's Study in Child Health. The National Human Genome Research Institute would have been the natural home. This one, though, Collins kept close, placing it in his own Office of the Director. What took shape is what you'd expect from the history that gave rise to it. The fundamental vision for the Precision Medicine Initiative's large cohort was the very same vision that the National Children's Study leadership resisted when Collins's request to piggyback revealed itself to be something else; it took on its prototypical form in the American Gene Environment Study in 2004; the groundwork for it was laid with the collaboration between the NIH and Pfizer at the heart of the Genetic Association Information Network; and it was facilitated by the work done with hospitals nationwide as part of the eMERGE Network that linked up genetic information with health data in electronic medical records.

The NIH, fearing that many Americans wouldn't know what a "cohort" was, eventually renamed the Precision Medicine Initiative Cohort Program the "All of Us Research Program" in 2016, but that rebranding didn't change the fact that it was still very much an exercise focused primarily upon genomic medicine. An advisory group endorsed the idea of recruiting the million participants with a convenience sample, which would be a tremendous resource for finding very small and very rare genetic effects. A Genomics Working Group convened in 2017 to assess the trajectory of the program and then issued a report that guided the plans for data collection, data analysis, and what genetic results to return to the participants. The program was described as a data-gathering platform, untethered to any specific hypotheses. Genomic Centers were announced in 2018 to prepare for receiving the participants' DNA and then processing it. The program prepared to join forces with a genetic counseling service that could work with participants to interpret their genetic results when they were returned.

The language surrounding All of Us also reflected this history. It incorporated all the marketing phrases as well as the conceptual confusion of the evolution from pharmacogenomics to personalized medicine to precision medicine during the previous two decades. The president's own language in the State of the Union address revealed this; it was the "Precision Medicine Initiative," but it delivered "personalized information." The FDA, which partnered with the NIH to roll out the program, described the effort to usher in an era of "precision medicine, sometimes known as 'personalized medicine.'" Traditional medicine, Collins told reporters covering the development of the program, was stuck in a one-size-fits-all mold; precision medicine, on the other hand, was about getting the right treatment to the right patient at the right time. The All of Us Research Program website characterized precision medicine as being all about treating "you as an individual." Familiar words all.

The National Children's Study, though officially terminated just a few weeks prior to the president's announcement in his State of the Union address, lurked ever present in the shadows of All of Us. What was once an object of national pride and scientific excitement became recharacterized as a billion-dollar object lesson in failure. How, journalists asked Collins, would the NIH "avoid the ghosts of its past" as it embarked on this new nationwide program? Collins and other designers of the All of Us Research Program went to great lengths to make it clear that the faults of the National Children's Study would not corrupt their new initiative. Collins's effort to replace the environmental study of kids with his own genomic design while Alexander led it, along with the study's disintegration after Collins took over the NIH, was never mentioned. Instead, blame was placed on the early leadership. That effort in Child Health, the official narrative went, was mismanaged; it struggled to keep up with an evolving scientific and technological landscape; it tried to do too many things at once. The new initiative in the Office of the Director, Collins promised, would not repeat those mistakes when it rolled out across the country in 2018.

8

Disparity in the Genome

ON THE AFTERNOON of May 6, 2018, a nationwide block party of sorts unfolded in seven cities across the United States. At Railroad Park in Birmingham, Alabama, students from Miles College, one of the nation's proud historically Black colleges and universities, treated onlookers to the gospel classic "Worship Him." The Pasco High School Mariachi Band played through a drizzling rain in Pasco, Washington, while dancers with Ballet Folklórico spun in dresses of brilliant vermilion, lime, and turquoise. Ford Field, home

Members of the Miles College Choir perform at the All of Us Research Program Launch Day celebration in Birmingham, Alabama, on May 6, 2018.
Photograph by Reginald D. Allen

of the Detroit Lions, was temporarily taken over by the local radio personality JoAnne Purtan, who gave away zoo passes and tickets to Detroit Tigers games. The Northwestern Medical Orchestra, made up of medical students, performed an eclectic program that ranged from Beethoven to *Indiana Jones* for spectators at Chicago's Millennium Park, while community members took pictures of themselves and added them to a mural designed by local artists in Nashville and Kansas City. And in Harlem, guests took their seats in the pews of the historic Abyssinian Baptist Church, one of the oldest African American Baptist churches in the country.

The performances, speakers, and events at each location had an unmistakable focus on local community, culture, diversity, and inclusivity. That was no accident. Communities of color have faced appalling health disparities for centuries in America, and the residents of those seven cities in many ways exemplified the medical injustices that still scarred a nation. Dick Durbin, the senior senator from Illinois, asked the Chicagoans to imagine getting on the city's Blue Line train: "If you take three stops on that Blue Line as you're heading west, you will see a disparity in life expectancy of nineteen years." An official from the Detroit Health Department told the audience on Ford Field about the shockingly high rates of infant mortality in their city, equivalent to many underdeveloped nations. A vice president from the University of Alabama, Birmingham, explained how Black Americans in the Deep South suffered from rates of heart disease and diabetes that far exceeded national averages. Dara Richardson-Heron, a community engagement expert, lamented to the congregation in Abyssinian how "health disparities are well-known but not at all well understood."

The gathering in cities across the country on that spring day was presented as a concerted effort to combat that problem. It was both a celebration and a call to action. The federal government had committed billions of taxpayer dollars to creating the largest medical research platform in the nation's history. The All of Us Research Program was aiming to enroll at least a million Americans. It was expected to reach a diverse population in all fifty states and was intended to last a decade and hopefully longer. From each of those million participants, the audiences were told, data would be collected on their unique lifestyle and nutrition, the particular environments in which they lived and moved and worked, and their individual biology. "No more one-size-fits-all," the National Institutes of Health director, Francis Collins, promised the Abyssinian congregation, "but optimized for each particular person. That's a concept we now call 'precision medicine.'"

This was where the call to action came in. The health disparities that existed in communities of color across America, the argument went, could only be faced head-on if the people from those communities went to a website titled "Precision Medicine—Prevent Health Disparities" and signed up to join All of Us. That was no small ask. Calvin O. Butts III, the long-serving reverend of Abyssinian Baptist Church, paced back and forth in front of the congregation and spoke in a voice that fluctuated between measured whispers and booming declarations. "There are a lot of reasons why we should not trust what's being presented here today." The history of research on untreated syphilis among Black American men involved in the Tuskegee Syphilis Study, the eugenic sterilization of Black American women without their consent, and the theft of cells from the body of Henrietta Lacks cast a dark shadow over what was supposed to be a joyous affair. "All of these things frighten us," Reverend Butts warned, "and they send signals that say, 'We do not want to be experimented on again.'" Scientific representatives from the All of Us Research Program, spread out across the seven sites, acknowledged that American biomedical research had a dismal history when it came to communities of color. But, they countered, this time was different. To the audience gathered in New York City, Collins said, "We can and we will do better, thanks to All of Us."

The assurances along with the commitment to focus the program on addressing health disparities were enough to marshal prominent leaders from those communities to step forward and vouch for the importance of participation. Reverend Butts, after giving voice to the concerns, spoke to the revolutionary potential of the new research program rolling out, saying, "Now I'm convinced All of Us means 'all of us.'" The president of the Delta Research and Educational Foundation, which promoted work that benefits Black American women and their families, told the attendees in Kansas City that "precision medicine is a very good thing" and that Black Americans, in order to take advantage, needed to "get on board the research train!" The president of the National Alliance for Hispanic Health, speaking in Spanish, implored the Hispanic residents of Pasco to enroll and be included. The civil rights icon and congressman from Illinois Bobby Rush led the Millennium Park crowd in a call-and-response chant: "All of Us! All of Us! Not Some of Us! But All of Us!" And even a great-granddaughter of Henrietta Lacks's took the stage in Chicago to recall the first time she heard about precision medicine's ability to swap out one-size-fits-all treatments with those tailored to the individual, hailing the moment when "we now have a seat at the table."

There was, however, a deep tension simmering beneath the celebrations. Precision medicine was described to the audiences in Nashville, Pasco, and Birmingham as producing health care tailored to the individual, based on detailed examinations of a patient's particular lifestyle, unique environment, and specific biology. The design of the All of Us Research Program, however, was skewed scientifically, financially, organizationally, and educationally toward DNA. No environmental working group or lifestyle working group convened alongside the Genomics Working Group as part of the early planning. The Genomics Working Group issued its strategic report to shape the basic design of All of Us; no environmental or lifestyle report was generated to guide the design decisions about sample size, sample distribution, and the necessity of hypotheses. The series of Genome Centers were funded to sequence the participants' DNA; no environmental or lifestyle centers were created in anticipation of the program's rollout. Genetic counseling services were recruited to assist participants with interpreting the genetic results that they received as part of their involvement; no such professional services were incorporated to help the participants navigate their dietary issues or exposure to environmental toxins. The massive enrollment number was ideal for a huge genomics study, but it would also mean little in the way of resources for carefully measuring environmental data from one million people. The plan to equip participants with environmental sensors that they wore to collect personal environmental exposure data was relegated to an aspiration for the future of the program, just as it was more than a decade earlier in the American Gene Environment Study. Environmental and lifestyle information was going to come instead from surveys and Fitbits, which could capture some self-reported information and data about heart rates or step counts, but were ill-suited to provide evidence related to environmental exposures or social phenomena about which a person may be entirely unaware. David Schwartz from the NIH's Environmental Health Institute warned Collins back in 2004 that the *E* in his AGES was underdeveloped; a decade later, the name of the program had changed from AGES to All of Us, but that fundamental fault remained the same.

The problem wasn't just that the attention to environmental and lifestyle information that was talked up in descriptions of precision medicine was played down in the actual design of the program. It was also that the genetics that was played up in the design was simultaneously muffled in the recruitment materials. The "All of Us Anthem," a video introducing the project, was shown at launch celebrations. It included soaring orchestral music, slow-pan shots of baseball fields and pastures and firefighters, and a velvety voice-over that reminded Americans, "When called upon to give from within, we come together and find that our capacity to help others is limitless." But it did not include a single reference to "genes," "genetics," or "DNA." Flyers, postcards, posters, and other recruitment items distributed to the public in towns and cities across the nation had their own omissions. References to "what's inside us," "blood," "family history," and "biology" offered only coded signals to what was at the heart of All of Us.

Can a massive study focused primarily on genetics narrow the nineteen-year gap in life expectancy between three stops on Chicago's Blue Line or reduce the frightening infant mortality in Detroit? Can prioritizing study of the DNA of Black Americans in the Deep South reveal why they are dying at higher rates of heart disease and diabetes? Two stories from the history of American biomedical research suggest that we should be profoundly skeptical of such expectations and profoundly worried about what happens when research participants who were promised one thing learn later that they were enrolled in something else entirely.

Why join the *All of Us* Research Program?

It may be that your family has a history of cancer, diabetes, or high blood pressure. Whatever the reason, you can help change the future of medicine by participating in the *All of Us* Research Program.

The *All of Us* Research Program is a historic effort to accelerate research and improve health. By taking into account individual differences in lifestyle, environment, and biology, researchers will uncover paths toward delivering personalized medicine.

The more we know about what makes people unique, the more customized health care can become. That is why we're asking one million or more people to join us.

You can schedule an appointment online by going to **JoinAllofUsPA.org/schedule**, or by calling **(800) 664-0480** or **(412) 383-2737**. Visit **JoinAllofUsPA.org** for more information.

*All participants receive $25 after completion of their visit.

Call **(800) 664-0480** or visit **JoinAllofUsPA.org/locations** for a full list of our convenient enrollment centers.

***All participants will receive $25 after completion of their visit. To complete the visit, participants must create an account, give consent, agree to share their electronic health records, answer health surveys, and have their measurements taken (height, weight, blood pressure, etc.). Participants may also be asked to give blood and urine samples.**

Sample All of Us Research Program recruitment mailer, encouraging residents in Pittsburgh to join the project and contribute to personalized medicine. Note the lack of any explicit reference to the genetics that rested at the heart of the effort. National Institutes of Health's All of Us Research Program, a registered trademark of the U.S. Department of Health and Human Services

. . .

Scientists from the NIH's National Institute of Arthritis and Metabolic Diseases arrived in Phoenix in February 1963. They then headed about thirty miles south to Sacaton, the largest town among the Gila River Indian Community. The federal researchers were there to study rheumatoid arthritis among the Pima, the Indigenous people who occupied that region for centuries. In order to study the impact of different climates on the affliction, the scientists intended to collect data from the Pima who lived in the hot and dry Sonoran Desert and compare it with data collected from the Blackfeet of Montana who lived at the cold and semiarid base of the Rocky Mountains. As part of that investigation, the NIH scientists collected blood from both populations, and when they examined the blood sugar levels among their samples, they discovered something alarming. The Pima suffered from diabetes at a rate higher than scientists had ever before documented.

Calling the Sonoran Desert "hot and dry" really doesn't do it justice. It is blazing hot, reaching temperatures in excess of 120°F during the summer months, and it is bone dry because only a dozen or so inches of rain fall there each year. Desert tortoises, gray foxes, and mountain lions eke out an existence among the saguaro cactus, prickly pear, and creosote bush. The desert landscape is harsh, but as the Gila River cut across southern Arizona, it historically traced a lush green arc through the parched Sonora.

For more than a millennium, until the fifteenth century, the Hohokam people thrived across what is now the southwest United States and northwest Mexico. Archaeological evidence reveals that they constructed tremendous irrigation works that dispersed water from the Gila River for miles and turned the surrounding landscape into rich agricultural fields. The Hohokam grew beans and squash as well as cotton and tobacco.

The Spanish first reached the region in the sixteenth century, and by then the Hohokam were no more. In their place were the Akimel O'odham and the Tohono O'odham who traced their ancestry to the Hohokam. "O'odham" means "people"; the Tohono O'odham were the "Desert People," and the Akimel O'odham were the "River People." The Akimel O'odham, like the Hohokam before them, shaped the earth and utilized the Gila River to irrigate fertile farmland. They were also skilled artists and basket weavers. The Spanish referred to the O'odham as the "Pima," and that name was in turn adopted by English-speaking traders of the eighteenth century as well as nineteenth-century gold rushers traveling across the Sonoran Desert on their way to California. The United States, after taking control of the region following the Mexican-American War and then the Gadsden Purchase, established the first reservation in what is now the state of Arizona south of Phoenix in 1859, and that's where the NIH scientists traveled just over a hundred years later to find a diabetes epidemic ravaging the community.

Roughly nine out of every ten cases of diabetes are the product of what we now call type 2 diabetes. Normally, when you eat highly processed carbohydrates or sweets, glucose floods into your circulatory system. Your pancreas then produces insulin, which helps shepherd the glucose out of your blood and into your body's cells, where it can be used for energy. But this pancreatic balancing act is not indestructible. If your body is exposed to high sugar levels over and over again, then the pancreas can become incapable of producing sufficient insulin, or the insulin that it produces can become less effective, or the cells that are supposed to take up the glucose can lose that receptivity. This leads to excess glucose moving throughout your body, and that wreaks havoc. Heart disease, kidney disease, nerve damage, a weakened immune system—all of these things can follow from diabetes. If untreated, the disease leads to blindness, poorly healing wounds that can require amputations, and kidney failure.

Health-care workers who served the Gila River Indian Community knew that diabetes was devastating the local population in the 1950s. But the sheer magnitude of the problem wasn't completely clear until the NIH scientists began sampling the people more broadly. By their calculations, nearly half of the community over the age of thirty had diabetes, which was ten times the national average. That reality, combined with the fact that the people tended to spend their whole lives in the area and the population as a whole was fairly homogeneous from a genetic perspective, created an opportunity for the researchers. If they could set up a research center in the area, then the scientists could study diabetes in the Gila River Indian Community and track how it impacted the people across generations, hopefully understanding and even eliminating the medical scourge.

On June 13, 1966, scientists and administrators from the NIH mingled with the tribal leaders to celebrate the dedication of a new field studies clinic in Sacaton. A local band provided entertainment for the gathered crowd, much as the Miles College Choir and the Pasco High School Mariachi Band performed for the Birmingham and Pasco communities five decades later. And in another striking similarity, the NIH scientists used the festive environment to encourage the local Indigenous population to do their part by signing up and joining the historic effort.

James Neel, a human geneticist at the University of Michigan,

The Sacaton field studies clinic, in 1966, built to study the diabetes epidemic among members of the Gila River Indian Community. *NIH Record*

introduced a new concept just months before the NIH scientists first arrived in Phoenix, and it had a profound impact on the inhabitants of the Gila River Indian Community. Neel thought diabetes was an evolutionary puzzle. It posed serious health risks and an impediment to successful reproduction, yet more and more humans were living with it. Why would something so maladaptive be simultaneously so widespread? The answer, Neel proposed, came from looking to the deeper evolutionary history of humans. For the vast majority of time that people existed on Earth, they were hunter-gatherers, moving across deserts, savannas, and plains desperately looking for the next available meal. Humans, as a result, had a feast-or-famine relationship with food. In response to this caloric environment, Neel hypothesized that humans developed what he called a "thrifty genotype": they became genetically equipped with a metabolic ability to absorb and store energy for long periods of time until the next feast presented itself. Neel presumed the genetic architecture of such a thrifty genotype must be quite simple, probably just an underlying gene or two. The rising rates of diabetes in the twentieth century, on Neel's hypothesis, were the evolutionary by-product of a thrifty genotype suddenly deposited in modern society. The "blessings of civilization" surrounded people with abundant calories, and so the genetic mechanism that allowed humans to survive in the past when food was scarce struggled in the present where food was everywhere.

Neel, in 1962, offered up this possible explanation for the prevalence of diabetes in humans generally, not in any specific group of humans. But the thrifty genotype concept morphed almost immediately after Neel introduced it. Since diabetes was particularly high among Indigenous communities across the globe, researchers studying those populations embraced the idea that they were suffering from high rates of diabetes because they had a *particular* version of the thrifty genotype. Something universal had become something racialized. Scientists began linking up the thrifty genotype with high rates of diabetes among the Alabama-Coushatta tribe, the Seneca of New York, Indigenous Australians, and Polynesians, and the NIH scientists in Arizona made the same connection for the community there. "If a 'thrifty genotype' were to occur anywhere," they pronounced, "it is not surprising that it would be found in these people who have subsisted for about 2,000 years by irrigation farming in the desert where the availability of water, and hence food, was intermittent." Neel himself visited the Gila River Indian Community in 1965 to work with the NIH scientists and see his thrifty genotype concept in action.

The thrifty genotype and its association with the Gila River Indian Community went mainstream in 1998 when *The New Yorker* published Malcolm Gladwell's "The Pima Paradox." He was, in many ways, simply summarizing what the scientists had been saying about diabetes since Neel had introduced his concept. "The diet of the Pima is bad, but no worse than anyone else's diet," Gladwell surmised. That meant the answer to the "Pima paradox" had to rest somewhere within the decedents of the Hohokam, either in their genomes or in what was perceived to be their tendency toward a sedentary lifestyle.

Gladwell put his finger on the basic rationale for how the thrifty genotype is used to explain racial health disparities. First, assume the environments of the different groups are generally similar. Then turn attention to what's left that could explain the different health outcomes in the different groups—either their different genes or their different behaviors. The fundamental fault in that reasoning rested on the first premise. The environments of many communities of color are not at all similar to the environments of many white communities, and the story of the Pima after the U.S. government created the Gila River Indian Reservation makes plain just how dangerous that assumption is.

The Gila River started drying up soon after 1859. White settlers, farmers, and ranchers to the east of the Gila River Indian Community began diverting enormous amounts of water away from the Gila River, and that meant there was insufficient water for the Pima by the time the river reached them. This decimated their farms and their agricultural way of life. Federal water rights resolutions should have brought that life-giving resource back to the region, but it failed them. A horrific "Forty Years of Famine" took hold of the region from 1870 to 1910, when the community faced population-wide starvation.

A dam was eventually built on the Gila River in the 1920s about a hundred miles east of Sacaton, and the Pima were promised a fair share of water released from the reservoir for their farming. That promise was not kept. As a result, the government began providing the Gila River Indian Community with regular supplies of surplus commodity foods. This solved the threat of starvation, but it created a new threat. The Pima had for centuries grown accustomed to eating a diet of beans, squash, corn, and fish. The federal commodities they were then made to rely upon instead consisted of things like lard, refined sugar, white flour, canned meats, candy, salty snacks, and cheese; fresh produce wasn't included until the 1990s. If a Pima family wanted to supplement the federal commodities with fresh fruits and vegetables, it required a one-hour-round-trip drive.

A health survey of the Pima at the beginning of the twentieth century revealed a single case of diabetes among their people. By the 1960s, almost half of the adults had it. That was not paradoxical; there was never a "Pima paradox." It was as if the entire population had been placed on the *Super Size Me* diet for decades. What would have been paradoxical is if the Indigenous population miraculously didn't develop a diabetes epidemic in response to the awful nutrition imposed on them.

The extremely high rate of diabetes among the community did indeed provide the NIH researchers with a scientific treasure. "Much of what we now know about diabetes comes from the Pima," goes the tribute. The basic endocrinology of diabetes, the relationship between diabetes and obesity, the different types of diabetes, the progression from diabetes to kidney failure—these insights and many others relied in part on data that the Gila River Indian Community members provided the NIH scientists. Anybody treated for diabetes today is in some way indebted to them for their contribution.

The Pima, however, benefited little from their participation. The decades-long search for a thrifty genotype came up empty. Genetic linkage studies and then genome-wide association studies turned up different regions of the human genome associated with slightly higher risk of type 2 diabetes, but none of those discoveries translated into a genetic explanation for why the Pima had such high rates. All the while, the diabetes epidemic among the Gila River Indian Community ran rampant, the horrible dietary source of the problem remaining unchanged.

In 1998, ironically the same year that "The Pima Paradox" appeared, James Neel admitted that his thrifty genotype concept needed reconsideration. The genetics of diabetes was far more complicated than he and most other human geneticists assumed in the 1960s. His basic story about the shift from a hunter-gatherer lifestyle was oversimplified, he granted. "The term 'thrifty genotype' has served its purpose," he conceded, "overtaken by the growing complexity of modern genetic medicine." The studies of Pima genetics, Neel concluded, "give no support" to the idea that the high rates of diabetes among Indigenous populations "might be due simply to an ethnic predisposition."

Still, the damage was done. The widespread application of the thrifty genotype concept to racial health disparities among Indigenous populations diverted decades of scientific attention and resources away from environmental prevention and toward a biomedical cure that never came. At the same time, it reframed the narrative about racial health disparities in diabetes among Indigenous populations. Rather than focusing on the glaring environmental racism that lurked all around the Pima, scientists fixated on the thrifty genotype, directing scrutiny on something inside the Pima body—a unique gene that supposedly made them uniquely unlucky from an evolutionary perspective.

By the twenty-first century, the Gila River Indian Community became increasingly frustrated with the results of their five-decade commitment to the NIH's study. The scientists were publishing hundreds of papers about the basic biology of diabetes, but the Pima were left living with a disease that continued to devastate. The community collectively decided to end active participation in the study, shifting their own attention and resources toward scientific research that focused on prevention and public health.

Dara Richardson-Heron, at the All of Us launch event in Harlem, complained that health disparities were "well-known but not at all well understood." That statement was only half-true. What is well understood about those disparities is that they are caused almost entirely by things in marginalized communities' physical, chemical, biological, and social environments. Decades of research make clear that the Pima are not unique in this regard. We know now that exposure to toxic air, water, and buildings combines with exposure to structural racism in schools, workplaces, and the criminal justice system, and then with lack of exposure to nutritious food, economic opportunity, and quality health care. All of that kills people or shortens their lives. What Richardson-Heron got right is that there are many unanswered questions about how all this works and in turn leads to the specific racial health disparities that plague America. What else is disproportionately in (or absent from) the physical environments of people of color that contributes to differences in health outcomes? Which political mechanisms perpetuate that unjust distribution? In what ways does environmental racism operate differently in a large city like Atlanta compared with a small town like Saylorsburg, Pennsylvania? In what ways do these exposures and phenomena play out differently in different communities of color, or among members of the same community but at different socioeconomic levels?

There is, in short, much still to be learned about health disparities. There is much to be learned about why infants die at tragically high rates in Detroit, why neighborhoods in Chicago separated by mere miles watch their citizens live to different ages, and why Black American women and men across the South suffer from chronic conditions at rates unmatched by their white neighbors. There is little reason to believe, though, that a massive study focusing primarily on genetics—no matter how long it lasts, nor how much money it receives, nor how many people it recruits, nor how diverse are those participants—offers the best route for solving or even understanding those appalling problems.

The lesson of the Pima warns of the dangers that follow from looking for answers in the wrong places, the risks of thinking that problems caused by structural racism in society can be fixed by biomedical research that prioritizes the study of genomes. Which leads to the second important cautionary tale for All of Us, one about structural racism built into the biomedical research process itself.

Sickle-cell disease and cystic fibrosis are, in many ways, quite similar. Patients with either condition are afflicted with symptoms often starting in infancy and resulting in a life cut short. For sickle-cell disease, the process begins in the blood. Red blood cells are typically smooth and disk shaped. Every time your heart beats, your red blood cells course through your arteries and veins, bouncing off cellular walls and each other, carrying oxygen to all parts of your body. Red blood cells of people with sickle-cell disease, though, are stiff and crescent, like a sickle. Those cells cannot transport oxygen efficiently, die sooner, and create logjams in the circulatory system, clogging vessels that can cut off oxygen to regions of the body; when that happens, people suffer pain that can be excruciating. Pain crises may last hours or days. The sickle cells can also damage retinas. They can induce chronic fatigue because oxygen is not adequately moving through the body; cause infections; break down lung tissue; increase a person's risk of stroke, hypertension, and heart disease.

Cystic fibrosis is equally debilitating. Lungs, pancreas, and intestines secrete fluids that lubricate the surrounding tissue and keep the body running smoothly, much like the oil in an automobile with an internal combustion engine. Someone with cystic fibrosis, however, has certain proteins that struggle to move chloride in and out of cells. The body's fluids, as a result, become thick and gooey. The viscous mucus, like sickled red blood cells, clogs up. A patient with cystic fibrosis can develop intestinal blockages and constipation due to the congestion in their digestive system; this often goes hand in hand with insufficient weight gain and growth for children with the disease. Lungs are particularly vulnerable. Shortness of breath, wheezing, even coughing up blood are common symptoms of cystic fibrosis. As time goes on, the damage done by the sticky fluids increases; lungs become inflamed and infected, and eventually the lung tissue breaks down and causes respiratory failure.

Sickle-cell disease and cystic fibrosis are both classic examples of autosomal recessive conditions. Everybody has two copies of both the cystic fibrosis transmembrane conductance regulator gene and the hemoglobin subunit beta gene; you received one copy of each from your mother and one copy of each from your father. Most people have two fully functioning copies of both genes. It's possible, though, to have a mutation in one of the copies and not even know it because, with recessive conditions, people can be asymptomatic as long as they have one functional copy; these people are called "carriers" in genetic parlance. It's only when egg and sperm both deliver a dysfunctional copy of the gene that a child has the disease.

A diagnosis of sickle cell or cystic fibrosis often means a life consumed with keeping the disease at bay—for both the child and their family. So much revolves around managing care and preventing the next emergency, and so patients and their families are ever vigilant. That also means educating neighbors and teachers and, later in life, co-workers and friends so that others can understand the complications associated with the conditions as well as the threat of a looming crisis.

Sickle-cell disease is far more common than cystic fibrosis, both in the United States and around the globe. There are, in any given year, about 30,000 people living with cystic fibrosis in America, while there are nearly 100,000 people with sickle cell. When you move outside the United States, that difference only magnifies. Estimates of the global prevalence of cystic fibrosis range from 70,000 to 100,000. Sickle cell affects between 6 and 7 million people worldwide.

The Nobel Prize winner Linus Pauling and his colleagues pinpointed the molecular source of sickle cell arising in an abnormal form of hemoglobin in 1949. That was extremely early by molecular medicine standards, and sickle cell became labeled the "first molecular disease." The basic biochemical puzzles surrounding cystic fibrosis weren't unpacked completely until the 1960s.

Suppose for a moment that all you knew about sickle-cell disease and cystic fibrosis came in the paragraphs above: similar basic genetics, similar toll on patients and their families, different prevalence, different scientific histories. Which condition do you suspect has received more attention and resources from the biomedical research community over the years? If you guessed sickle cell, you'd be entirely within reason; all else being equal, it's the one that affects far more people and that was understood earlier. You'd also be wrong. That's because there's one more difference between sickle-cell disease and cystic fibrosis: sickle cell primarily affects people with African ancestry; cystic fibrosis primarily affects people with European ancestry.

It's obvious that translating scientific discoveries into health interventions requires resources. Identifying a gene associated with some disease doesn't automatically cure that disease; in the best-case scenario, it's the first step in a long process toward using that basic biological information to alter the molecular system in some way so as to offset the harm done by the genetic deficiency. That requires lots of money. John Strouse, a hematologist and sickle-cell disease expert, has documented over the years the enormous gap in financial support for sickle-cell patients compared with cystic fibrosis patients. A great deal of the research on and awareness of rare diseases like those two conditions is driven by private foundations that support the promising scientist, organize the walkathon fundraiser, keep families abreast of the latest medical advance. Foundation dollars often come from private donors, who are more likely to commit money to causes with which they are personally familiar. Since white people are more likely to come in contact with someone who has cystic fibrosis and, for all sorts of unjust historical reasons, have more financial resources than Black Americans, cystic fibrosis foundations have access to a larger pool of potential support even though there are fewer people living with the condition. The numbers aren't even close. Sickle-cell foundations generally have less than $10 million to spend annually promoting the interests of their patients and families; cystic fibrosis foundations' annual expenditures are in the hundreds of millions of dollars. Strouse also found disparities between how much the federal government spent on cystic fibrosis and sickle-cell research per patient.

Those differences in monetary support naturally translate into differences in research, treatment options, and outcomes for the two patient populations. In any given year, there are typically far more scientific publications devoted to cystic fibrosis than to sickle cell; the FDA has approved significantly more drugs for cystic fibrosis compared with sickle-cell disease; and there are many more comprehensive care facilities designed specifically for cystic fibrosis patients than there are for sickle-cell patients.

Nearly half a million people had signed up to join the All of Us Research Program by 2022, four years after the launch events in those seven cities across America. Half were from diverse races and ethnicities. Those people were told that a detailed look at their lifestyles, environments, and biologies would shape the future of biomedical research, even though the program is designed and oriented primarily around their genes. That's 500,000 people enticed by the opportunity to help prevent racial health disparities in their communities, even though there's little reason to believe that a scientific project focused on genes is the best way to address those problems. Five hundred thousand of the planned one million.

The most likely outcome of the All of Us Research Program for marginalized communities struggling with health disparities is that they will find themselves in the same frustrating situation as the Pima. Decades from now, they will see their data used to produce countless scientific publications. Scientists may even start saying things like "much of what we now know about heart disease comes from the All of Us participants." Those insights, however, will be of little help in providing solutions to racial health disparities because the main causes of the differences in health outcomes between racial groups aren't in our genomes. But imagine for a moment that the largest genetic effort in history identified genes that turn out to be responsible for infant mortality in Detroit or heart disease in Birmingham. What then? Well, people with those genes will be in roughly the same position as people with sickle-cell disease in 1949, when that "first molecular disease" was identified. They'll face a biomedical research landscape infused with structural racism, one that directs resources disproportionately toward genetic conditions that impact populations with a greater ability to pay for the development and use of medical interventions.

9

The "Gleevec Scenario"

CHRONIC MYELOGENOUS LEUKEMIA is a rare blood cancer that arises when a person's bone marrow produces excessive white blood cells. The disease develops slowly, so many people don't know that their body is malfunctioning for quite some time. Eventually, though, the overly abundant leukocytes disrupt the circulatory system, robbing it of space for the oxygen-carrying red blood cells and clogging up organs tasked with cleansing the system. Fatigue, weight loss, fever, and abdominal pain ensue. Prior to 2001, the prognosis for somebody diagnosed with chronic myelogenous leukemia—five thousand to ten thousand new cases in the United States each year—was bleak. Available treatments included interferon therapy, which tended to make people feel as if they were living with a horrific case of the flu, and a bone marrow transplant, which many people didn't survive. If a patient wasn't properly suitable for one of those treatments or if they'd already tried a treatment but the cancer returned, then the proliferating white blood cells ultimately overwhelmed the body, leading to infections, an enlarged spleen, uncontrolled bleeding, oxygen starvation, stroke-like symptoms, heart failure, and death. Life expectancy upon diagnosis was less than five years; few survived for ten.

That changed when the U.S. Food and Drug Administration approved imatinib (brand name Gleevec) in 2001. Most chronic myelogenous leukemia patients who can consistently take Gleevec live for five years; many survive far longer. Indeed, oncologists found that Gleevec turned their patients' leukemia into a chronic condition, something more akin to the long-term management of type 2 diabetes than the life-and-death struggle against a deadly cancer. Gleevec also had milder side effects than the previous treatments, which is why it was hailed upon its arrival as a "miracle cure" and a "magic bullet." It even played a role in episodes of the prime-time dramas *Law & Order* and *The West Wing.*

The added excitement surrounding Gleevec was that it represented a new way of practicing medicine. Gleevec works by targeting the protein made by a specific gene. It's that protein that triggers a molecular cascade in the bone marrow resulting in the excessive leukocytes. By inhibiting the protein, Gleevec shuts down the chronic myelogenous leukemia process at its source. Appearing on the national stage in 2001 just as the Human Genome Project was wrapping up and enthusiasm for personalized medicine was growing, Gleevec became the designated poster child for gene-guided medicine. "Before Gleevec" and "After Gleevec" became a handy way to mark the historic moment when health care transitioned into the age of the right treatment, for the right patient, at the right time. Advocates of genomics predicted that the "Gleevec scenario" would become the norm in medicine.

Two decades on, we can see that Gleevec genuinely altered health care—but not in the way the champions of genomic medicine predicted. Rather than ushering in a new era of widely accessible and broadly applicable miracle cures and magic pharmaceutical bullets, it ushered in a new era of drugs that don't work for most patients and are so exorbitantly priced that they risk bankrupting some of those who are biologically eligible. A closer look at the story of Gleevec provides a glimpse of the dangerous path that lies ahead for a health-care industry that places DNA at the center of medicine.

. . .

Brian Druker, an oncologist at Oregon Health and Science University in Portland, phoned the general information number of the FDA in 1996, asking if he could speak with somebody in the toxicology department. He hoped the conversation might resolve a disagreement between him and personnel at the drug giant Novartis over whether a chemical compound, what Novartis called STI-571, was ready to go to human trials. Druker thought the drug had tremendous potential for his patients with chronic myelogenous leukemia. He'd tested STI-571 in vitro and on mice and was thrilled to see how effectively it attacked cancerous cells in a targeted way. The Novartis leadership, however, were reluctant. They were worried about tentative signs of toxicity in experiments involving dogs, and so their plan was to proceed slowly with their drug, trying STI-571 on additional animal models first, which could delay bringing the product to market for months or years, or could mean the compound never reached human trials.

Druker has the lean, wiry build of a long-distance runner, along with the easy smile and slow timbre of his midwestern roots. He was attracted to the science of cancer as a young undergraduate, fascinated by the laboratory work involved in uncovering oncological mechanisms. That was then matched, in medical school and residency, with a passion for providing care to the sickest patients, often diagnosed with cancer. Watching his own patients succumb to chronic myelogenous leukemia, Druker thought Novartis was being overly cautious. He initially reached out to a handful of prominent experts in clinical trials, hoping that they'd be able to assess the situation for him. But Druker was relatively unknown in the community at that time, and nobody got back to him. So he phoned the FDA and explained what his research had shown. Would the regulatory agency take seriously an application to proceed with a human trial based on that work? The federal employee, to Druker's delight, told him that he and Novartis already had more impressive data than many companies asking the FDA for permission. Druker relayed the good news to Novartis, thinking they'd say, "Great!" Instead, he got an earful for ignoring normal procedure. Still, Druker's move created internal pressure at Novartis to overcome its reservations and see what STI-571 could do for leukemia patients harboring the Philadelphia chromosome.

Peter Nowell, a scientist at the University of Pennsylvania, and David Hungerford, a graduate student at the nearby Institute for Cancer Research in the Philadelphia suburbs, first noticed an unusual, small chromosome in patients with chronic myelogenous leukemia in the late 1950s. The discovery of that chromosomal abnormality in the City of Brotherly Love provided the name and forever linked it to the blood cancer. But how that cellular phenomenon related to the deadly cancer remained a mystery. Did the Philadelphia chromosome cause leukemia, or was it instead an effect of the cancer? The second piece of the puzzle came a decade later when Janet Rowley at the University of Chicago utilized new ways to stain and visualize chromosomes, revealing the Philadelphia chromosome to be the product of a rare translocation event wherein chromosome 9 and chromosome 22 swapped segments during cell division. Rowley's insight both explained the unusual Philadelphia chromosome that resulted from the translocation—the already tiny chromosome 22 gave away a portion of itself in exchange for an even smaller fragment from chromosome 9—and directed attention to the place where chromosomes 9 and 22 joined up as the potential site of the problem. Subsequent genetic research in the 1980s revealed that the segment of chromosome 9 latched onto 22 in such a way that a gene carried along for the ride was set in overdrive, producing an enzyme that stimulated uncontrolled cell growth in bone marrow and leading to the particular form of blood cancer.

Druker, armed with the accumulation of that scientific information, sought a chemical compound that could combat his patients' chronic myelogenous leukemia in a more humane way. Cancer treatment at the time, Druker recalled, was like trying to turn off the light in a room by blindly swinging a baseball bat around and hoping you connected with the lightbulb at some point; it was debilitating for patients because the chemotherapies smashed healthy and diseased cells alike. Druker wanted a treatment that chemically flipped off the light switch, one that homed in on the cancer and left the healthy cells alone.

Of the compounds that Druker tested, STI-571 was the best contender. It was a chemical that Ciba-Geigy (Novartis's precursor) initially developed as an anti-inflammatory but then left sitting idle. There was significant hesitancy at both Ciba-Geigy and then Novartis to devote resources toward developing a drug for chronic myelogenous leukemia because there were relatively few patients with the rare disease. It was the market segmentation worry that critics of pharmacogenomics raised throughout the 1990s: Why invest time and money developing a drug that treated just 15 percent of leukemia patients, most of whom didn't even live that long? Marketing analysts within Novartis warned the company leadership that the drug would probably make only about $100 million for the company. The promising data that Druker highlighted, though, suggested it really was worth investigating STI-571.

Druker and Novartis, with the green light from the FDA finally in hand, proceeded with the first clinical trial in 1998. STI-571 was administered to about thirty people who had already failed to respond to interferon therapy or found their cancer returned after that treatment, and were also ineligible for a bone marrow transplant. The results that came in just months later were like nothing oncologists had ever seen before. Every participant responded positively to the drug, with only moderate side effects.

That first trial took place during the nascent years of the internet, and rumors started circulating in online chat rooms about STI-571. Patients with chronic myelogenous leukemia began contacting Druker and asking their doctors to get them access to Novartis's drug. Druker, for his part, was caught off guard by the sudden demand because the results weren't even public yet, he told a *New York Times* reporter years later: "I'd never written it up. I hadn't presented the data. Their doctors thought I was a charlatan." STI-571 seemed too good to be true. Sensing the excitement, Druker hoped Novartis would finally get behind the compound, but once again the company balked, expressing the same objections. When Suzan McNamara, a patient with the leukemia, contacted Druker in 1999 and begged to be enrolled in a study so that she could get access to STI-571, Druker encouraged her to communicate with Novartis leadership directly. Perhaps the voice of a patient would carry more weight. McNamara in fact launched a website—no small task at a time when dial-up internet was slow and search engines were still in their infancy—inviting patients with the disease to sign on and join a request for expanded access to the lifesaving experimental treatment. Thousands joined McNamara, and when her letter reached the desk of Novartis's CEO, Daniel Vasella, he finally agreed to prioritize STI-571.

Over the next two years, the drug was given to thousands of patients. The results were genuinely astounding: more than 90 percent of the participants saw their white blood count return to normal. Novartis submitted its application to the FDA in early 2001. Ten weeks later the agency approved imatinib, and Gleevec was introduced to the world.

Politicians and reporters alike flocked to the excitement surrounding Gleevec. Vasella joined Tommy Thompson, President George W. Bush's first secretary of health and human services, at a press conference where Thompson explained how the drug sailed through the FDA approval process in record time due to its enormous promise. Imatinib was described variously as a "smart bomb" and a "miracle cure." The tiny pills appeared on the cover of *Time* magazine under the lines "There is new ammunition in the war against cancer. These are the bullets." The messaging all beat the same drum: Gleevec heralded a new future in oncology specifically and health care generally where treatments were tailored to patients.

The one sign of concern regarding imatinib in 2001, however, was its price. The article in *Time,* after conveying the enthusiasm, concluded by pointing out that the drug sold for well over $2,000 a month. That translated into an annual cost of between $25,000 and $30,000. Vasella admitted the price was steep, but there was good reason, he countered. The existing interferon therapy was priced similarly. Moreover, the market for imatinib was small, and Novartis typically invested $600–$800 million in the research and development of a new drug. The higher-than-usual price tag was necessary to offset the company's financial commitment to the lifesaving treatment; Novartis might not even make much of a profit on its new drug. Vasella said that the price could come down if the population of patients who took the drug expanded.

The price of imatinib stayed in the $25,000–$30,000-a-year range for some time. Then there was a development that foreshadowed the future of health-care economics guided by the pursuit of genomic medicine. Around 2006, the price began climbing. The market for Gleevec had indeed expanded after its release because it changed a deadly blood cancer into something that could be managed like a chronic disease, which by Vasella's own reasoning should have brought the price down. Around that same time several other drugs that worked similarly, called "tyrosine kinase inhibitors" based on the rogue proteins that they shut down, also came out alongside imatinib. One, in fact, was Novartis's own Tasigna. These newcomers were priced even higher than Gleevec, in the $5,000–$7,000-a-month range. So the listed price of Gleevec gravitated up toward that of the new arrivals. Starting late in the first decade of the twenty-first century, the price of imatinib steadily rose, 5 percent one year, 8 percent another, then nearly 20 percent. Patients who paid $2,200 a month in 2001 were paying triple that amount a decade later. By 2011, Novartis was making more than $4 billion annually from the drug that Druker had to beg the company to produce and prioritize. Recouping the pharmaceutical company's R&D investment clearly was no longer a viable explanation for the economics of imatinib.

The chronic myelogenous leukemia community, appalled by what was unfolding, eventually cried foul. In 2012, patients and loved ones of those with the disease posted a petition to Change.org asking members of Congress to intervene on Novartis's price hikes. One patient who was in Druker's original trial of STI-571 complained, "It is borderline criminal to force people to make the choice between life (being able to afford the Gleevec) and death (being financially unable to buy the drug that will save their lives)." Another supporter of the petition simply pleaded, "My grandma needs this medicine." More than a hundred chronic myelogenous leukemia physicians and researchers also took to the fight, penning an article for the premier hematological journal *Blood* that sounded an alarm about the "unsustainable prices of cancer drugs." Gleevec, they warned, was setting a dangerous precedent that other cancer drugs were chasing, which was normalizing profiteering in the drug industry and driving up the costs of health care. Druker was initially hesitant about signing on to the editorial; as a practicing oncologist and researcher, he was dependent on maintaining a good working relationship with the drug industry. But he ultimately agreed to add his name after a friend and the leader of the effort asked him, "Brian, if we don't do this, who will?"

The rising price of Gleevec was what garnered most of the critical scrutiny, but Novartis was also employing other tactics to protect and even extend its monopoly on imatinib. The patent was set to expire in 2013, but the company filed for a series of extensions which moved that deadline to 2015. Then, when the patent was up, Novartis utilized a pay-for-delay strategy, the details of which were hidden behind the veil of a confidentiality settlement, paying off the manufacturer of the first approved generic to withhold distributing its imatinib for seven months, thereby delaying the arrival of a competitor until 2016. All the while, Novartis marketing gradually directed attention toward Tasigna, saying it was even better than Gleevec, which would encourage oncologists to switch patients to a Novartis product that had years left on its patent protections.

The normal rules of economics clearly no longer applied. That troublesome picture only became clearer when generic imatinib was released in 2016. With conventional drug markets, the arrival of generics led to significant price drops because competition drove them down. Generic imatinib, however, arrived in the United States with a list price of about $5,000 a month, far more than what Gleevec had first fetched in 2001. The science advertised as personalized medicine played by its own market rules.

It wasn't until 2019, when a number of generic versions of imatinib had arrived, that the price of the lifesaving treatment for chronic myelogenous leukemia finally dipped to a reasonable rate. Novartis, by that point, had made roughly $50 billion in global sales of Gleevec. Meanwhile, the cost of treating chronic myelogenous leukemia had plateaued around 2015–16 when the Gleevec patent expired and the generic version of imatinib came online. Physicians had indeed moved their patients over to one of the newer tyrosine kinase inhibitors like Tasigna. Chronic myelogenous leukemia, as a result, continues to be an extremely pricey diagnosis twenty years after Gleevec appeared on the market, driven in large part by the steady arrival of shockingly expensive drugs that follow the financial example set by Novartis and its magic bullet.

. . .

Erin Havel was familiar with the routine surrounding monthly treatments for her vascular malformation. Those rare circulatory phenomena are the product of an abnormal development in blood vessels, usually present from birth. They can range from small and innocuous to large and life threatening. Havel's began as a shoe-shaped discoloration on her side. Nagging back and neck pain in her twenties led her to visit a physician who performed an MRI, which revealed spiderwebs of aberrant vessels, some about her torso and back, others through her neck. The treatment that Havel received involved monthly injections of an ethanol compound into the spindly extensions to chemically cauterize them. She traveled to Denver each month to meet her vascular malformation specialist. After standard blood work, she was placed under general anesthesia, injected with the ethanol, and then given a day to recover from the medically induced hangover that followed from the alcohol exposure.

As Havel went through the standard pre-procedure questions on a visit in 2007, the physician's assistant said, "We're not going to do your treatment today." When Havel asked why, she was told that her white blood cell count was unusually high. "It could either be a virus," the physician's assistant replied, "or leukemia." It was leukemia.

Havel had just turned thirty. Her passion was performance—theater, singing, playing the piano and guitar. She was hardworking and mindful of her finances. Now the already complicated process of being treated for her vascular malformation was joined by talk of Philadelphia chromosomes, forcing her to consult a leukemia specialist. The cancer diagnosis terrified Havel; she was comforted, however, by what her oncologist had to say about a drug named Gleevec. She'd have to take the pills for the rest of her life, but if she could stay on the medication, it was likely she would die with leukemia, not from it.

"Thank God I had access to health insurance," Havel thought when the first prescription for Gleevec arrived. Gleevec, to Havel's shock, would have cost her $3,800 a month if she wasn't insured.

The diagnosis of leukemia came at a particularly vulnerable moment for Havel. She and her partner were in the process of moving to a new city to follow the partner's job; that left Havel unemployed and reliant on her partner's employer-provided health insurance. "I've found that illness either cements a relationship or rips it apart," Havel recalled. In their case, it was the latter. That forced Havel to find another source of health insurance of her own, but the coverage plans that she identified and for which she qualified made accessing the pricey Gleevec extremely difficult. Medicare, by her calculations, provided her only enough to cover one month's worth of her prescription for imatinib. Supplemental plans demanded that she pay thousands of dollars up front before she could even get her first Gleevec prescription filled, and then she would have to absorb a percentage of the price thereafter. Novartis's drug, all the while, kept getting more expensive. Havel watched it go up $1,000 the first year, then another $1,000 the next. Soon it was rising every month. When an opportunity arose for her to attend a meeting in Philadelphia for chronic myelogenous leukemia patients hosted by another manufacturer of tyrosine kinase inhibitors, she asked her oncologist if there was any message he wanted relayed. He laughed at first but then paused, adding sincerely, "Actually, if you could mention the costs of the [tyrosine kinase inhibitors] and the problems that go along with getting the medication for people, that would be good."

The normally money-wise Havel, even with health insurance, eventually found her savings depleted; she'd already accumulated tens of thousands of dollars in credit card debt tied to her medical expenses. Havel's reliance on Gleevec was indefinite. Any gap in taking those pills, her doctor warned her, put her health at risk. As her debt grew, Havel contacted an attorney to assess her situation: Would she have to declare bankruptcy? To her disappointment and embarrassment, the attorney told Havel that she effectively already was bankrupt. It was just a matter of filling out the paperwork.

Havel's trip to Philadelphia gave her the opportunity to meet a whole community of chronic myelogenous leukemia patients. She learned from them that her financial challenges and anxieties surrounding Gleevec were by no means unique. Over lunch one afternoon, attendees shared stories of "drug smuggling"; if their physicians moved them from one medication to another, they'd save their remaining pills from the old prescription and find a patient who needed them but couldn't afford them. It wasn't legal, but the infraction seemed far less important than providing the life-or-death pills to someone who struggled to get a prescription filled. The patients that Havel spoke to in Philadelphia represented a microcosm of the chronic myelogenous leukemia community; many people suffering from the disease don't get their prescriptions filled on schedule within just months of diagnosis, which, as Havel's oncologist cautioned, translated into grave health risks.

Druker and the other chronic myelogenous leukemia experts warned in 2013 that Gleevec was setting an unsustainable example in health care. They were right. In the two decades that followed the arrival of imatinib, the costs of oncological therapies increased significantly. Treatments operating under both the "personalized medicine" and the "precision medicine" labels were by no means solely responsible for that rise, but they were a major driver. New precision medicines in oncology have a median price of $150,000 a year, the same annual price that Gleevec reached before its first generic arrived. On top of that, any drug that matches only certain molecular-genetic profiles needs a diagnostic test to go along with it. Those tests can cost hundreds or even thousands of dollars each. A recent survey of oncologists found that more than two-thirds of them predicted the expanding reliance on precision medicines meant higher costs for the future of cancer care.

Personalized medicine was first marketed in the late 1990s on the promise that it would reduce the dangerous side effects associated with drug toxicity. As the earlier story of MaryAnne DiCanto attested, its performance on that front has been mixed. It did, however, introduce another side effect: financial toxicity. Oncologists, alarmed by the growing prevalence of cases like Havel's, have started warning their patients about the economic implications of cancer care. One study found that more than 40 percent of cancer patients exhaust their entire life's assets within two years of diagnosis. The diagnosis also makes somebody more than two and a half times more likely to go bankrupt, a lesson Havel learned firsthand. Right when a person needs to focus their energy on staying healthy and battling cancer, they're hit with the additional anxiety associated with trying to stay financially afloat.

The threat of financial toxicity for any given patient navigating the health care industry in America relies on so many different variables: the presence or absence of health insurance, judgments that insurance companies make about what treatments and diagnostics are or aren't covered, judgments that insurance companies make about what percentages of covered treatments and diagnostics a patient must shoulder, the availability of financial assistance programs offered by a pharmaceutical company, whether or not a patient qualifies for such programs, whether or not a patient can navigate the convoluted bureaucratic process required to apply for such programs, the duration of their illness, the duration of their treatment, whether or not they can stay employed during illness and treatment, the person's wealth and financial and social support available from family and friends, even where they live. A patient has little to no control over most of those things, which means it's very hard to predict who will come out of a cancer diagnosis solvent and who will be financially devastated by the experience. Scott Primiano, DiCanto's husband, estimated that they spent about half a million dollars in out-of-pocket expenses on her cancer care. They were fortunate to have had good health insurance and to have been relatively well-off.

Kristen Kilmer, who like DiCanto was diagnosed with breast cancer, didn't fare so well. Kilmer also underwent genomic testing to identify drug matches. In her case, AstraZeneca's Lynparza was singled out as a fit for her genetic mutation. That was the good news. The bad news was that it was priced at $17,000 a month, and her health insurance would cover only a fraction of that amount, deeming it experimental for her particular case. Kilmer applied to financial assistance programs, which provided her access to treatments for some time. But then she was suddenly judged to no longer qualify. Kilmer had already spent $80,000 in out-of-pocket expenses on her cancer care. Like Havel, she was also thousands of dollars in credit card debt. Kilmer was presented with a tragic decision to make: forgo treatment, knowing it numbered the remaining days with her family, or seek treatment and ultimately leave her husband and daughter in financial ruin. Kilmer shared her heartbreaking plight with Liz Szabo, the same Kaiser Health News journalist who took DiCanto's story to national newspapers, explaining to Szabo that she'd ultimately opted to stop taking the Lynparza. "It's not worth it," she told the journalist. "I will not put my family into that kind of debt." Hours after Szabo's story went live, AstraZeneca notified Kilmer that she qualified again for the financial assistance program. Her Lynparza was in the mail.

Havel, fortunately, eventually landed on her feet. She met someone new. Long dreaming of having a child, Havel worked with her oncologist to plan a safe pregnancy and delivery all while monitoring her chronic myelogenous leukemia. She now has good, reliable health insurance that allows her to fill a monthly prescription of generic imatinib for just $5. When generic imatinib came online and observers were forecasting a sudden drop in the price of the drug, Havel warned in an essay for a cancer patient advocacy website that the generic could very well be priced nearly as high as brand-name Gleevec. And when a subsequent oncologist encouraged Havel to switch to one of the newer, expensive tyrosine kinase inhibitors, Havel said, no thanks, she was doing just fine on the imatinib. Still, Havel keeps a handful of the pills stockpiled, just in case imatinib throws one more financial surprise at her.

. . .

Before the All of Us Research Program rolled out across the United States in 2018, before President Obama's 2015 State of the Union address, before the National Children's Study was terminated, before the collaboration with Pfizer, the American Gene Environment Study, and even the American Family Study—before all of that, Francis Collins took to the stage at the Smithsonian National Museum of Natural History in April 2003 to celebrate the completion of the Human Genome Project. And just before he took out his guitar to sing about "D-D-D-D-DNA," he predicted that a revolution was coming to health care, one where genomics would identify basic biological defects and then tailor prediction, prevention, and treatment plans to individual patients. "And if you think that sounds like science fiction," he said, "I'll tell you about everybody's favorite poster child for this particular paradigm." Collins told the crowd about the history of the Philadelphia chromosome and Brian Druker's work with Novartis's STI-571. He recounted the initial clinical trial that transpired five years earlier and its thrilling results. Gleevec, in 2003, was in its honeymoon phase—FDA approval, widespread public excitement about the new "magic bullets," no sign yet of the pricing controversies to come. "And we want to see many more Gleevecs," the future director of the NIH said, beaming.

The heralded arrival of Gleevec provided a tangible example of genomic medicine in action. Collins's predictions, in turn, were premised on the idea that the Gleevec story could be generalized, that other diseases would play by similar biological rules as chronic myelogenous leukemia and so follow a similar path to gene-guided interventions. In five to ten years, Collins forecast, the "major contributing genes" for common illnesses like diabetes, asthma, and Parkinson's would be discovered in the human genome. That paved the way for individualized, preventive medicine where, by 2010, audience members would be able to find out what was "lurking in your future on the basis of a predictive genetic test." Finally, by 2020, "the Gleevec scenario will have begun to play out in a most gratifying way." By that point, Collins saw a health-care landscape where imatinib was the norm, "where we will have a gene-based designer drug available for almost any disease that you can name because we will have uncovered the molecular basis based on this genomic revolution."

The vision to which Collins treated his audience was certainly optimistic, but he wasn't operating outside the general excitement about the future that was common in the genetics community in 2003. After all, Gleevec was indicative of a biomedical research industry where a major genetic swerve was already under way. Pharmaceutical companies like Pfizer and Novartis had gradually set aside their concerns about market segmentation and embraced pharmacogenomics as a tool for drug discovery, drug testing, and drug prescribing. Geneticists, initially worried about the limits of their family-based linkage studies to find new genes, were rejuvenated by Neil Risch and Kathleen Merikangas's advice to shift to larger and larger genome-wide association studies aimed at finding genes with smaller effects. Biotechnology companies like Illumina were rolling out a lineup of new machines that made obtaining genetic information faster and cheaper. Both policy makers and reporters were smitten with talk of "personalized medicine," an approach to health care that promised to treat each patient as an individual rather than with a defunct one-size-fits-all approach. It seemed as if all the pieces were in place to go find the genes for diseases like diabetes, asthma, and Parkinson's and then let the "Gleevec scenario" play out for them too.

That's not what happened. The geneticists, instead, quickly bumped into their missing heritability problem. Rather than finding the common genes associated with the common illnesses that had large enough effects to provide reliable routes to prediction and intervention for the general population, the geneticists found regions in the human genome that were either very rare or common but very limited in terms of the impact that they made on the most widespread health problems. Studies of Parkinson's disease, for example, gravitated toward two genes: one called leucine-rich repeat kinase 2 (*LRRK2*), and the other called the glucocerebrosidase (*GBA*) gene. *LRRK2* was implicated in some familial cases of Parkinson's disease, but familial Parkinson's makes up only a small fraction of the illness. In the more general population, *LRRK2* was associated with only a few percent of patients with the condition. *GBA*, for its part, fell into the more common but limited category. The risky variant of that gene appears with more frequency than *LRRK2*, but it conveys only moderate risk. When the strongest genetic predictors of Parkinson's are combined, they account for only 5–10 percent of all the cases of that illness. Asthma genetics followed a similar trajectory. Genome-wide association studies quickly fixated on a cluster of genes found on chromosome 17, but they accounted for only a very small portion of the difference in who developed asthma and who did not. The genetics of diabetes proved insightful for some very rare forms of that disease, but for the most common type—type 2 diabetes, which makes up about 90 percent of all diabetes cases—hundreds of single nucleotide polymorphisms were picked up which each contribute very little to the prevalence of that illness in society. The picture that emerged within just five years of the completion of the Human Genome Project was that for the most common diseases that were taking the biggest toll on society, the "major contributing genes" just weren't that major. There simply weren't any Philadelphia chromosome equivalents for those traits to be found.

The widespread utility of genetic prediction and genetic intervention was premised on the assumption that the major genes would indeed be major. So when that assumption proved erroneous, the clinical applications that were to follow also became compromised. Almost all people with Parkinson's don't carry the risky *GBA* or *LRRK2* variants, and most people with those risky variants don't develop Parkinson's. Asthma prediction based on somebody's genetic profile is equally limited. The genome-wide association studies on asthma turned up dozens of locations in the human genome associated with the respiratory illness, but the incremental risk that each conveyed was so small that clinical prediction wasn't feasible. For type 2 diabetes, the common single nucleotide polymorphisms discovered were also individually limited. One way around that problem, geneticists hoped, was to combine all the information together into a composite polygenic risk score, one that projected the danger for somebody based on the collective risk spread out across their whole genome. That, it turned out, did offer some predictive ability for diabetes. The problem, though, was that the predictive value of the genetic information was inferior to the predictive value of other information that doctors already routinely collected about their patients, things like body mass index and blood sugar levels. So even where genes did get some predictive purchase, they were outmatched by the standard things that physicians had long been doing without adding the costs of genetics into the mix. For most people and most diseases, a predictive genetic test is a very unreliable way to judge what's "lurking in your future." That was true in 2010, and it remains true today.

Gene-guided drug development has had mixed results. It was certainly not the case that a "gene-based designer drug" became available for almost any disease. Illnesses like asthma and type 2 diabetes resisted personalized medicine drug discovery precisely because the genes that were identified in the genome-wide association studies provided so little in the way of viable drug targets. Still, the ease with which DNA could be gathered from patients continued to attract industry-sponsored science. The genomic approach made strides in places where subgroups of people with some disease could be partitioned off based on their rare genes. Pharmaceutical companies like Pfizer, Biogen, and GSK are all in the process of testing specific compounds tailored to Parkinson's patients with the *LRRK2* gene. Sanofi and others have drugs in clinical trials geared toward patients with the *GBA* variant. Private industry is investing hundreds of millions of dollars in treatments that are designed to work for a small percentage of Parkinson's patients. If any of those contenders ultimately reach FDA approval, history suggests that they'll follow the Gleevec pricing paradigm, adding Parkinson's disease to the list of illnesses where physicians must warn their patients about the risk of financial toxicity.

Despite the tremendous excitement surrounding the promises of personalized and then precision medicine, despite the impressive technological advances that have made obtaining genetic information fast and cheap, despite the enormous financial investment committed to making gene-based health care a reality, and despite the bipartisan embrace of genomic medicine, the actual number of people who can benefit from genetically informed patient care is severely limited. That limit creates problems for both the excluded and the included. For most people suffering from the vast majority of diseases, information about their DNA offers very little in the way of useful guidance toward prediction, prevention, and treatment; and yet, because of all the financial, technological, political, and media pressures documented in these pages, they will still be encouraged to add their DNA to one or more of the proliferating genetic datasets. For those to whom it does apply, it presents an economic landscape imperiled by financial ruin; patients with rare diseases or patients with rare forms of common diseases are made reliant on private industry, where the primary focus on profit ensures that any market segmentation created by underlying biology will be met with price increases to offset that limited population of consumers.

Throughout the late twentieth and into the early twenty-first centuries, as health care in America was lurching toward genetics, something frightening was simultaneously unfolding. The rates of diseases like Parkinson's, asthma, and diabetes were all increasing. Phrases like the "Parkinson's pandemic" and the "asthma epidemic" became used routinely to characterize trends in health outcomes that were spiraling dangerously out of control. Parkinson's disease is among the fastest-growing neurological disorders in the world; the number of people diagnosed with it doubled between 1990 and 2015, and the United States carries a particularly heavy burden. About 1 million people were estimated to have Parkinson's in the United States in 2020, and that number is expected to grow to 1.6 million by 2040. Asthma rates also spiked in America throughout the 1980s and into the 1990s. Then, between 1999 and 2018, the prevalence of asthma rose from about 9 percent to 13 percent across the nation. The toll was particularly heavy on Black children, whose asthma rates went up 50 percent in the first decade of the twenty-first century. Diabetes in America followed a similar pattern: rapid increases in the 1980s and 1990s, followed by a shallower climb after the turn of the century. Again, though, that hid a more alarming trend within communities of color. The percentage of white Americans with diabetes is half what it is in Native American populations. New diagnoses of type 2 diabetes in Black youth doubled between 2003 and 2015, a statistic nearly matched by Hispanic children.

There are known physical, chemical, biological, and social features of the environment that contribute to all three of these common diseases, where differences in both individual and group exposures account for differences in health outcomes. Parkinson's disease was virtually unknown prior to the Industrial Revolution, and rates of Parkinson's are increasing fastest where industrialization expands quickest. Industrial solvents, lead, air pollution, polychlorinated biphenyls, and pesticides are all linked to the neurological disorder. Increasing rates of asthma have been tied to contaminants in homes and schools, toxins in the air, and exposures in the workplace. Both low birth weight and obesity later in life also increase risk of developing asthma. Rising rates of obesity are adding to the diabetes epidemic too, especially in kids. Children are particularly vulnerable to the metabolic harm done by the proliferation of highly processed foods. Exposure to mothers' gestational diabetes while in the womb is another danger. The built environments of neighborhoods—their walkability, the availability of green space, proximity to healthy foods, even noise levels—all figure into diabetes risk.

The threats posed by unsafe living environments are clear, and yet there are still many ongoing debates about the nature of those problems and how to effectively intervene in order to limit the harm they do. How do pesticides make their way into the human body and contribute to Parkinson's risk—in drinking water, on produce, some other way? Rachel Carson flagged the concern sixty years ago. The danger persists, only the attention has shifted from DDT to substances like glyphosate, used in herbicides, and organophosphates, common in insecticides. To what extent are increases in rates of certain diseases a function of an aging population, and what, in turn, is it about aging that increases risk in some but not other conditions? Asthma was long thought to be the result of an increasingly sanitary environment, the thought being that children were growing up in more hygienic worlds, thereby weakening their immune systems; this "hygiene hypothesis," however, struggled to make sense of the profound racial health disparities in asthma because children of color are more often exposed to less hygienic spaces. Why are rates of diabetes rising so quickly in children generally and children of color specifically? How does exposure to gestational diabetes while in utero increase risk for a child later in life?

One thing is certain: the rising cases and the gaping racial health disparities are not the result of genes. The Pima made clear decades ago that no thrifty genotype was awaiting discovery in the genome-wide association studies that would explain the gap between white and Native American rates of that debilitating disease. The genes of Black children in the United States did not evolve over the course of just one decade to account for a 50 percent increase in asthma. Six hundred thousand more Americans aren't going to develop Parkinson's disease by 2040 due to a sudden change in gene frequencies.

The way to begin addressing those population health problems is to understand them, and the best way to understand them is by studying the actual determinants of health in our environments. That research won't be as politically popular as genetic research promising to personalize medicine or make health care more precise. It won't hold out the promise of major profits for private industry. It will require cooperation and leadership from federal, state, and local governments, as well as trust from communities willing to put their faith in those institutions. It may very well be costly and time consuming. And the research won't generate headlines about "miracle cures." But if the goal is to reverse the negative health trends that are imperiling a nation, if it is to more equitably improve the health of the whole population, if it is to reduce the unjust burden of racial health disparities, if it is to keep people healthy and out of hospitals rather than seeking expensive treatments in hospitals, and if it is to make health a shared goal rather than an individual burden, then it's the only route available to us.

Acknowledgments

The research for this book was based in part on conversations that I had with many scientists (affiliated with the federal government, universities, hospitals, and private industry), patients and their family members, elected officials, presidential appointees, journalists, federal staff and administrators, pharmaceutical executives, lobbyists, staff at nonprofits, hospital executives, and budget officials, some of whom asked not to be named; I am grateful for their time and their recollections of the various episodes in which they played a part: Edward Abrahams, Keith Albert, Russ Altman, David Balshaw, Bill Barnes, Louise Bier, Vence Bonham, Jeff Botkin, Carol Browner, Steve Buka, Connie Citro, Edward Clark, Adolfo Correa, Dana Dabelea, Stephanie Devaney, Todd Dickinson, Brian Druker, Greg Duncan, Barbara Entwisle, Alan Fleischman, Stephen Galli, Eric Green, Jeff Gruen, Joe Grzymski, Alan Guttmacher, Glen Hanson, Erin Havel, Lee Hood, Dora Hughes, Jason Hwang, Charlie Johnson, Fujio Kayama, Sarah Keim, Woodie Kessel, Raynard Kington, Mark Klebanoff, Philip Landrigan, Bruce Lanphear, Mike Leavitt, Claude Lenfant, Martin Mackay, Teri Manolio, Don Mattison, Robert Michael, Marie Lynn Miranda, Jeff Murray, Maynard Olson, Nigel Paneth, Philip Pizzo, John Porter, Steve Prescott, Samuel Preston, Scott Primiano, Bill Riley, Neil Risch, Laura Lyman Rodriguez, Gualberto Ruaño, David Savitz, Hillary Savoie, Catherine Schaefer, Peter Scheidt, David Schwartz, Albert Seymour, Ellen Silbergeld, Bonny Specker, Anne Spence, Shannon Stitzel, Liz Szabo, John Thompson, Chris Vaccaro, David Walt, Marc Williams, Marshalyn Yeargin-Allsopp, and Elias Zerhouni.

From that larger group, a handful are deserving of particular recognition. Edward Clark, Alan Guttmacher, Sarah Keim, Nigel Paneth, and Peter Scheidt all spoke with me on numerous occasions and shared documents and correspondence from the period. Those extended discussions and primary sources were truly invaluable.

I also benefited from being given access to primary sources at the National Human Genome Research Institute by Christopher Donohue and his wonderful team in the History of Genomics Program there. The impetus for this project occurred when they kindly invited me to participate in a workshop at the Genome Institute in April 2015, where I was among the early group of scholars who were introduced to the incredible trove of documents that they had archived and were digitally processing for historical research. Coincidentally, that meeting took place not long after President Obama's 2015 State of the Union address, during which he announced what would become the All of Us Research Program. It was while I was reading through items in the archive and saw the seeds of All of Us planted more than a decade earlier that I first thought there might be an interesting story worth telling.

A number of colleagues, mentors, friends, and students read all or portions of earlier drafts of the manuscript and provided me with constructive feedback. I shudder to think of what the book might have looked like without the input of Jeff Botkin, Teneille Brown, Luca Brunelli, Jorge Contreras, Carl Craver, Thomas Cunningham, Stephen Downes, Joyce Havstad, Lucas Matthews, Erik Parens, Lisa Parker, Chris Phillips, Rich Purcell, Cliff Rosky, Ken Schaffner, Kathryn Tabb, Eric Turkheimer, Joseph Yracheta, students in Steve Downes's course on genetic causation at the University of Utah, students in Jon Fuller's course on philosophy of medicine at the University of Pittsburgh, and participants in Rachel Dentinger's history and philosophy of science reading group at the University of Utah.

Several people were instrumental in taking what was initially a fairly narrow project about different types of research at one federal agency and turning it into a broader reflection on reasons to be alarmed about the direction of biomedical research generally. John Knight and Jordan Fisher Smith both gave me tremendous advice on the book proposal, challenging me to "open the aperture" and show

how the lessons of the story reached beyond the halls and labs of the National Institutes of Health; John also read a penultimate version of the manuscript and took a personal interest in thoughtfully advising me on some of the subtler aspects of story craft.

Michelle Tessler, my literary agent, masterfully helped this author new to the world of trade publishing navigate those intimidating waters; always encouraging, Michelle answered each question, advised on any matter, and anticipated every contingency.

And I am honored to call Jon Segal at Knopf my editor. He saw the potential for what the book could be from our first conversation, as well as the dynamic at the heart of it, and he patiently guided me along the path toward realizing that vision. There are very few sentences on these pages that weren't touched by Jon's generous pencil; I am indebted to him for all the work that he put into the manuscript and into making me a better writer. Jon's team at Knopf and, in particular, Sarah Perrin helpfully led me through the process of turning that manuscript into the pages of this book.

The book is dedicated to my wife, Dawn-Marie, for everything that she gave to me and it over the eight years that the project unfolded. She memorized all the acronyms, learned all the names, matched up all the affiliations, and kept track of all the plotlines, all in the service of helping me pull together what was often a jumble of thoughts and shape them into something coherent and convincing. She was a sounding board, a reader, an editor, one of the experts whom I consulted on matters of public school education, and the thing for which I am most grateful—a constant source of support.

Notes

INTRODUCTION

3 It is one of the most common cancers: Statistics on lung cancer are monitored and updated by the Centers for Disease Control and Prevention at www.cdc.gov. The American Cancer Society is another useful resource for data regarding lung cancer deaths (and its comparison to other cancer deaths): www.cancer.org.

4 Lung cancer is so deadly: Helmet 2016; Herbst, Morgensztern, and Boshoff 2018. For a philosophical perspective on how oncologists study and explain cancer, see Plutynski 2018.

5 Erlotinib, for example, was initially approved: M. H. Cohen, Johnson, Chen, et al. 2005; Speake, Holloway, and Costello 2005.

5 A study published just months before: Zhou, Wu, Chen, et al. 2011.

5 "a novel paradigm": Ginsburg and Phillips 2018, 696.

5 "revolutionary": Akdis and Ballas 2016, 1359.

5 "a new era": Collins and Varmus 2015, 793.

5 Efforts are under way to apply the model: Fernandes, Williams, Steiner, et al. 2017; Divaris 2017; Konishi 2017; Filipp 2018; Straatsma 2018; D'Ambrosia 2018.

5 a more affordable alternative: Wechsler 2018; Glorikian 2014; Jain 2019; Harvard Business Review Analytic Services 2018.

6 combating unjust racial and socioeconomic: Dankwa-Mullan, Bull, and Sy 2015; Landry, Ali, Williams, et al. 2018; Armstrong 2017.

7 personalized medicines themselves are exorbitantly expensive: Chase 2020. GoodRx keeps tabs on the most expensive drugs. Its top ten in 2020 included Zolgensma at No. 1, used to treat the very rare spinal muscular atrophy; of the nine others, seven were for rare genetic diseases and two were for cancers.

7 "patients with lung cancer who have the epidermal": Herbst, Morgensztern, and Boshoff 2018.

7 the more it succeeds at individualization: My father was prescribed erlotinib in 2011 and 2012, before a generic version of it was approved and available. One might think that the high costs of personalized medicines are only problematic when the drugs are patent protected, and that the prices inevitably drop precipitously once

the generics hit the market. However, as we'll see in chapter 9, that does not seem to be the case with personalized medicines; see, for example, Tessema, Kesselheim, and Sinha 2020 for an introduction to this economic phenomenon.

7 patient spending on their health care: Bradley, Yabroff, Mariotta, et al. 2017.

7 translated into decreased rates of survival: Goulart, Unger, Chennupati, et al. 2021.

7 financial toxicity: Gilligan, Alberts, Roe, et al. 2018; Ramsey, Blough, Kirchhoff, et al. 2013; Remon, Bonastre, and Besse 2016. Liz Szabo, a health reporter for Kaiser Health News, has documented in agonizing detail the toll this genomic oncology takes on families—the hype, the disappointment, and the financial ruin; see, for example, Szabo 2018a and 2018b.

8 performed less often on Black patients: Bach, Cramer, Warren, et al. 1999.

8 A 2019 study revealed: Kehl, Lathan, Johnson, et al. 2019.

9 Erlotinib and drugs like it: Tartarone, Lazzari, Lerose, et al. 2013.

9 the actual benefit typically means: Marquart, Chen, and Prasad 2018; Prasad 2020.

10 caused primarily by exposure: Proctor 2012; Brawley, Glynn, Khuri, et al. 2013; Lantz, Mendez, and Philbert 2013. Epidemiologists have known at least since the 1950s that occupational exposure to asbestos increases risk of lung cancer; see Doll 1955.

10 In the Columbus, Ohio, neighborhood: My father grew up in the 43026 zip code of Columbus; radon levels across Ohio can be found here: www.city-data.com.

10 The region sits adjacent: Berrett 2007; Hanley 1986.

10 Benjamin Franklin: Franklin 1735. Franklin wrote, "In the first place, as *an Ounce of Prevention is worth a Pound of Cure,* I would advise 'em to take Care how they suffer living Brands-ends, or Coals in a full Shovel, to be carried out of one Room into another, or up or down Stairs, unless in a Warmingpan shut; for Scraps of Fire may fall into Chinks, and make no Appearance till Midnight; when your Stairs being in flames, you may be forced, (as I once was) to leap out of your Windows, and hazard your Necks to avoid being oven-roasted."

11 rates of lung cancer in the United States: De Groot, Wu, Carter, et al. 2018.

11 Health disparities, it is abundantly clear: See Keisha Ray's *Black Health* (Forthcoming) for an introduction to the literature on the social determinants of health and their impact on health inequities.

11 Black patients with lung cancer: Ibid.

11 aren't novel observations: In the chapters to come where I tell the story of how personalized medicine first took shape, I'll introduce the criticisms that ensued. There are also more recent concerns raised regarding personalized and precision medicine; Dickenson 2013 was an early and powerful critique. Nigel Paneth, Michael Joyner, Richard Cooper, Sandro Galea, and Ronald Bayer have been the most consistent and forceful critics over the last decade in Bayer and Galea 2015, Joyner 2015, Joyner and Paneth 2015, Chowkwanyun, Bayer, and Galea 2018, Joyner and Paneth 2019a, Joyner and Paneth 2019b, Cooper and Paneth 2020a, Cooper and Paneth 2020b.

12 In 2000, you could count: Data on the increase in these drugs approved by the FDA is regularly reported by the Personalized Medicine Coalition in its *Personal-*

ized Medicine Report: Opportunities, Challenges, and the Future, the last version of which was published in Personalized Medicine Coalition 2020.

12 in 2018, two out of every five: Personalized Medicine Coalition 2018.

12 a pharmaceutical industry that is shifting: Das 2017.

12 The president and CEO of Geisinger: Geisinger's David T. Feinberg announced the plan to sequence all patients in Geisinger's system at the HLTH Conference in Las Vegas in May 2018: www.geisinger.org. Geisinger's collaboration with Regeneron began in 2014: www.genomeweb.com. The union, however, didn't receive a great deal of scrutiny until Feinberg announced the plan to sequence every Geisinger patient several years later: www.clinicalomics.com.

12 All of Us Research Program: President Barack Obama made the first public announcement of what would become the All of Us Research Program in his 2015 State of the Union address: "Remarks by the President on Precision Medicine," obamawhitehouse.archives.gov. That was soon followed by the NIH's announcement. The story behind why the All of Us Research Program is skewed so heavily toward genetics will become clear over the course of this book.

13 forces that are driving this embrace: There are similarities between, on the one hand, the dynamic that I am describing here about the contrast between investment in gene-centric personalized medicine and investment in more environmentally focused health research, and, on the other hand, the contrast between investment in biomedical/clinical medicine and public/population health. On that divide, see for example Krieger 2011; Löwy 2011; and Valles 2018 and 2020. My goal in this book is to show how that biomedical/clinical-public/population divide maps onto a nature-nurture divide, which sets the foundation for the kind of research that gets conducted. It's also worth mentioning here the fact that the book is admittedly focused on health research in the United States. That is partly about design; as will become clear in the pages to come, the main story that runs through this book took place at a U.S. federal agency (the National Institutes of Health), and so the political, economic, and social forces that impacted it were predominantly American. But it's also about the necessity of constraining the scope of a book. The dynamic documented here may very well be found in other nations; however, the presence of a universal health-care system, a stronger history of public investment in social services, a more limited role for pharmaceutical companies in the health-care industry, a greater appreciation for the value of environmental research and protection, and a less narcissistic and more community-oriented ethos would all presumably alter the dynamic significantly.

14 "book of life": Pennisi 2000, 2304; Karow 2001.

14 "holy grail of biology": Meek and Ellison 2000.

14 As Francis Collins, director: National Human Genome Research Institute 2003.

CHAPTER 1 A TALE OF TWO REVOLUTIONS

16 Olga Owens Huckins and her husband: Prince 2012; Knox 2012.

17 she wrote a letter: A copy of Huckins's letter can be found here: brbl-zoom.library

.yale.edu. The description of what she and her husband saw on their property after the spraying comes from this letter.

17 DDT was hailed: Kindela 2013.

17 Carson began communicating: Lear 1997; Lytle 2007; Souder 2012.

18 *Silent Spring:* R. Carson 1962. The working title of the book was *Man Against the Earth,* and "Silent Spring" was just the title of the chapter on birds, but the powerful auditory message of that chapter came to embody the whole book.

18 Carson's critics: M. B. Smith 2001; Sideris and Moore 2007.

19 galvanized the modern American: Dunlap and Mertig 1992; Gottlieb 1993; Sellers 2012.

19 Santa Barbara oil spill: Spezio 2018.

19 igniting the river in flames: Stradling and Stradling 2008.

19 In 1972, the EPA banned: Of note, DDT may still be administered in the United States in cases of public health emergencies.

19 Rachel Carson is remembered: Hynes 1985; Montrie 2018.

20 "now stored in the bodies": Carson discusses the health impacts of pesticides in chapter 3 of *Silent Spring* ("Elixirs of Death").

20 For millennia, public health: Porter 2005; S. Jones 2006; G. Rosen (1958) 2015. Rosen's history is particularly insightful here because he points out that, though rare, there was occasional attention to the public health threats posed by humans' own activities; for example, restrictions were placed on tanners and dyers washing or dumping their products in waterways that provided drinking water to a community. G. Rosen (1958) 2015, 21–22.

20 Toxic waste buried in the 1940s and 1950s: Colton and Skinner 1996; Newman 2016.

20 In 1982, Black Americans: McGurty 2007.

22 Subsequent research revealed: U.S. General Accounting Office 1983.

22 "environmental racism": Bullard 1990 and 1993; J. T. Roberts and Toffolon-Weiss 2001; D. Taylor 2014; Zimring 2015; Valles 2018; Washington 2019; Ray 2021.

22 But pediatricians cautioned that exposure: National Academies 1993.

23 National Children's Study: The Children's Health Act called for the large, longitudinal study, but it didn't give it a name. For some time after, the study leadership toyed with various names: "The Longitudinal Cohort Study of Effects on Child Health and Development"; "SEARCH: Study of Environmental Aspects Regarding Child Health." Eventually, they opted for brevity and went with "The National Children's Study."

23 Arno Motulsky: A. G. Motulsky 2016; H. Motulsky and Jarvik 2018.

24 Division of Medical Genetics: Comfort 2012.

24 Motulsky published a paper: A. G. Motulsky 1957.

24 Such research would never be approved: Hornblum 1997; Comfort 2009; Miller 2013.

24 In 1945, *Life* magazine published an essay: *Life* 1945.

24 Starting in the 1950s: Alving, Craige, Pullman, et al. 1948; Hockwald, Arnold, Clayman, et al. 1952; P. E. Carson, Flanagan, Ickes, et al. 1956.

26 the father of pharmacogenetics: Roehr 2018; Offord 2018. Motulsky typically gets credit for being the father of pharmacogenetics, but there were earlier scientists who also pointed to the relationship between someone's biological makeup and their response to treatment. William Osler 1892 warned of "great variability" among patients when it came to diagnosis and treatment, suggesting medicine was more art than science. And Alfred Baring Garrod 1909 suggested genetics could impact drug metabolism via a person's "chemical individuality." Motulsky, in contrast to Osler and Garrod, wrote at a time when examples of pharmacogenetics started appearing in the published literature, allowing him to take a hypothetical discussion and make it empirically concrete.

26 Friedrich Vogel: Vogel 1959.

26 Werner Kalow's 1962 book, *Pharmacogenetics*: Kalow 1962.

26 an editorial and an article about the event: *The New York Times* 1962; Schmeck 1962. For scholarly reflections on this history, see D. S. Jones 2013, Perlman and Govindaraju 2016.

26 "a new way to find out": Dreifus 2008.

27 one-pill-fits-all business model: Li 2014.

27 Cytochrome P450 encompasses: Idle and Smith 1995; Miners and Birkett 1998; Weber 2001; Danielson 2002. Daniel Nebert and Frank Gonzalez, both at the National Institutes of Health at the time, were instrumental in putting the molecular genetics of cytochrome P450 on the map for pharmacogeneticists. See, for example, Nebert, Adesnik, Coon, et al. 1987; Nebert and Gonzalez 1987; Gonzalez, Skodat, Kimura, et al. 1988; Nebert and Russell 2002.

27 Dietary factors: Bailey, Malcolm, Arnold, et al. 1998; R. Z. Harris, Jang, and Tsunoda 2003.

28 double-helical structure of deoxyribonucleic acid: Watson and Crick 1953. See Olby 1974 for the history leading up to the Watson-Crick discovery.

28 A gene is a segment: There is an enormous philosophical literature on the concept of the gene—its history, its various meanings, its evolution; Griffiths and Stotz 2013 is a good place to start for wading into that scholarship.

28 the terminology shifted to "pharmacogenomics": Hedgecoe 2003; Goldstein, Tate, and Sisodiya 2003.

28 adverse drug reactions: Abraham 1995. The alarming drug deaths study can be found in Lazarou, Pomeranz, and Corey 1998. There was debate about whether this study overestimated the number of deaths; see, for example, Grady 1998. Still, pharmaceutical companies, pharmacogenomicists, drug regulators, and the press picked up on the study and reiterated its shocking results, so the problem of adverse drug reactions swamped the debate about the specific numbers.

28 "pharmacogenomics" became a common term: Housman and Ledley 1998; Hedgecoe and Martin 2003; Blankstein 2014.

29 The SNP Consortium: Wellcome Trust Press Office 1999; A. L. Holden 2002.

29 Gualberto Ruaño: Hathaway 1997; Ruaño and Kidd 1992; Ruaño, Deinard, Tishkoff, et al. 1994.

29 Genaissance Pharmaceuticals: Oestreicher 2002.

30 "Pioneering Personalized Medicine": Ruaño interview, Sept. 23, 2019. Andrew Marshall, an editor at *Nature Biotechnology* at the time, was the first to report on the phrase "personalized medicine" with regard to its pharmacogenomics meaning, and his references link to conversations he had with Ruaño: A. Marshall 1997a and 1997b. Hathaway 1997 reports that Ruaño was clearly using the phrase as early as the spring of 1997 at the public launch of Genaissance, so it is likely that Ruaño coined the phrase. Of course, it's also possible that "personalized medicine" was floating around in conversations with pharmaceutical executives, and Ruaño was just the first to brand it, but I found no evidence of an earlier usage than Ruaño's, let alone an earlier published reference. For more on the emergence of "personalized medicine" in Marshall's publications, see Michl 2015. British sociologists of science have done extensive work probing the rise of personalized medicine out of pharmacogenomics in the late 1990s; see, for example, Hedgecoe 2004, Tutton 2012, and Tutton and Jamie 2013.

30 science journalists and editors: Hathaway 1997; A. Marshall 1997a and 1997b; Stix 1998; K. F. Schmidt 1998; A. Marshall 1998; Langreth, Waldholz, and Moore 1999; Langreth and Waldholz 1999; Connor 1999; Kolata 1999.

31 April 2003 completion of the Human Genome Project: *Bringing the Genome to You* 2003. A video recording of the event is available at videocast.nih.gov. Watson attended in person. Crick, just a year from death, addressed the audience with a prerecorded video.

32 When Collins sang: Collins declined or ignored multiple requests to participate in interviews about the episodes discussed in this book.

32 Collins grew up: Collins recounts his childhood in chapter 1 of his autobiographical Collins 2006.

34 Collins made three in five years: Riordan, Rommens, Kerem, et al. 1989; Wallace, Marchuk, Andersen, et al. 1990; Huntington's Disease Collaborative Research Group 1993.

34 "moments of worship": PBS 1998.

34 a shake-up unfolded at the NIH: Cook-Deegan 1995, 326–40.

34 Collins was hesitant, but he prayed: Collins 2006, chap. 5.

35 transition the National Center: Cook-Deegan 1995, 326–40.

35 Collins received a phone call: Shreeve 2004, chap. 1.

36 Reporters ate up the competition: Thompson 1999; Wade 1999.

36 Collins navigated the pressure: K. Davies, 2001; Sulston and Ferry 2002; Shreeve 2004.

37 The crowd gathered to officially declare: National Human Genome Research Institute 2000. The Human Genome Project was predicted to cost $3 billion and finish in 2005; instead, it cost $2.7 billion and was completed in 2003.

38 By 2003, a number of other countries: Ollier, Sprosen, and Peakman 2005; Metspalu 2004.

39 "We haven't had much discussion": Collins email, May 8, 2003. Collins's team at the Genome Institute put together a very brief draft of what such a study could look like, under the heading "US Health Population Study." 7/23/2003 Duane Alex-

ander, Director NIDCD, AG Large Cohort Genotype-Phenotype: Large Cohort Study, Draft Proposal (June 16, 2003), Box 4 of 4 Deputy Director Files, Folder 14, NHGRI History of Genomics Program Archive, Bethesda, MD.

39 Duane Alexander: Mills 1998.

39 Robert Cooke: Snyder 2014.

42 Despite approving the study: National Children's Study Interagency Coordinating Committee 2003.

43 Child Health began preparing a presentation: Sarah Keim notes, March 23, April 2, July 7, 2003.

43 The whole conversation lasted: Scheidt notes, July 23, 2003; Scheidt interviews, Sept. 25 and Oct. 6, 2015.

43 After the Genome leadership: Scheidt interview, Oct. 6, 2015.

43 Collins, back at the Genome Institute: Collins email, Aug. 19, 2003, 9/9/2003 The National Children's study White House Briefing: Emails about large scale cohort study, Box 4 of 4 Deputy Director Files, Folder 19, NHGRI History of Genomics Program Archive, Bethesda, MD.

44 The property sat on Rockville Pike: Lyons 2006.

44 Zerhouni, nominated by President Bush: Zerhouni interview, Jan. 20, 2022.

45 The American Family Study: The American Family Study was an abbreviation of the longer American Family Health and Environment Study; The American Family Health and Environment Study, White House Briefing, Sept. 9, 2003, Box 030, Folder 25, NHGRI History of Genomics Program Archive, Bethesda, MD.

44 Zerhouni, Collins, and Alexander presented the material: Margaret Cousley email, Nov. 18, 2003, 11/25/2003 American Family Health Study: Emails about Meeting with Margaret Spellings, Box 4 of 4 Deputy Director Files, Folder 27, NHGRI History of Genomics Program Archive, Bethesda, MD.

45 they organized a scientific gathering: Workshop on a Proposed American Family Study: Agenda, 12/1–3/2003 Large Cohort Meeting: Agenda, Box 4 of 4 Deputy Director Files, Folder 28, NHGRI History of Genomics Program Archive, Bethesda, MD; Workshop on a Proposed American Family Study: Participants, 12/1–3/2003 Large Cohort Meeting: Participants, Box 4 of 4 Deputy Director Files, Folder 28, NHGRI History of Genomics Program Archive, Bethesda, MD; Workshop on a Proposed American Family Study: Overview, 12/1–3/2003 Large Cohort Meeting: Workshop on a Proposed American Family Study: Overview, Box 4 of 4 Deputy Director Files, Folder 28, NHGRI History of Genomics Program Archive, Bethesda, MD.

45 The participants identified a number of challenges: Collins notes from Workshop on a Proposed American Family Study, 12/1–3/2003 Large Cohort Meeting: Handwritten notes (Dec. 1–3, 2003), Box 4 of 4 Deputy Director Files, Folder 28, NHGRI History of Genomics Program Archive, Bethesda, MD.

46 all signals indicated the collaboration: Workshop on a Proposed American Family Study: Meeting Summary, AGES Cohort Study 2/04–4/04: Workshop on a proposed American Family Study 12/1–3/2003, Box 4 of 4 Deputy Director Files, Folder 33, NHGRI History of Genomics Program Archive, Bethesda, MD.

46 announce his support for NASA: NASA 2004.

46 none of those issues were insurmountable: Collins email, Nov. 9, 2003, American Family Study 11/03–12/2/03: Emails about specific discussion items for AFS working group, Box 4 of 4 Deputy Director Files, Folder 21, NHGRI History of Genomics Program Archive, Bethesda, MD.

47 When he visited the White House: The American Family Health and Environment Study, White House Briefing (Nov. 25, 2003), 11/25/2003 American Family Health Study: American Family Health Study PowerPoint, Box 4 of 4 Deputy Director Files, Folder 27, NHGRI History of Genomics Program Archive, Bethesda. MD.

47 And when the geneticists crafted text: "American Family Health and Environment Study: Draft Remarks for a Possible Presidential Message," 11/25/2003 American Family Health Study: American Family Health Study draft remarks for a possible presidential message, Box 4 of 4 Deputy Director Files, Folder 27, NHGRI History of Genomics Program Archive, Bethesda, MD.

47 A cautionary proverb circulates: I encountered a number of spins on this proverb about Collins. The merger/acquisition phrase. "Francis is the only person who can walk in a revolving door behind you and come out ahead." Others that are better left unprinted. Due to the personal and sensitive nature of these comments, I am not providing my specific sources.

48 Guttmacher updated Collins: Guttmacher email, Jan. 14, 2004, AFHES 12/3/2003–1/31/2004: Email from AG to FC about NCS, Box 4 of 4 Deputy Director Files, Folder 29, NHGRI History of Genomics Program Archive, Bethesda, MD.

48 Scheidt conveyed a similar message: Scheidt email, Jan. 14, 2004.

CHAPTER 2 THE END OF A PARTNERSHIP

50 He was dying: Woolner 2017.

50 America in the mid-twentieth century: Levy and Brink 2005.

50 With that goal in mind: Mahmood, Levy, Vasan, et al. 2013. The Framingham Heart Study is now jointly run by the NIH and scientists at Boston University.

51 The Framingham study forever changed: Wong and Levy 2013; D. S. Jones and Oppenheimer 2017.

51 The heart study, though: The Framingham study leaders did eventually add a racially and ethnically diverse group to the study—the OMNI cohort. But that didn't come until 1994. Tsao and Vasan 2015.

51 As the postwar baby boom peaked: Klebanoff 2009.

52 Like the Framingham study: The National Institute of Neurological Diseases and Blindness (also created under Harry Truman's presidency) became the National Institute of Neurological Diseases and Stroke in 1968.

52 Janet Hardy: Blackburn 2005; F. N. Rasmussen 2008; Schudel 2008.

52 high-quality, objectively measured data: Hardy, Drage, and Jackson 1979.

53 Hardy was obsessive about staying in touch: Hardy 2003.

53 but it proved well worth it: Klebanoff 2009.

54 "Well, it's been forty years": Marshalyn Yeargin-Allsopp, an epidemiologist at the CDC, said this. Yeargin-Allsopp interview, Feb. 1, 2016; Mark Klebanoff interview, Jan. 5, 2016.

54 The Talmud, for example, warns: F. Rosner 1969.

54 In sacred Hindu texts: Corcos 1984.

55 The Hippocratic Corpus: Hippocrates 1923, *The Sacred Disease,* sec. 5.

55 The discipline of genetics: Olby 1985; Darden 1991; Kohler 1994; Fox Keller 2002.

55 These eugenicists feared: Scholarly research on the history of eugenics really exploded in the 1990s and then the first decade of the twenty-first century; a sample of the most influential works includes Paul 1995, C. Rosen 2004, Kline 2005, Stern 2005, Schoen 2005, Lombardo 2008, and Largent 2011.

56 Fortunately, eugenics gradually lost its scientific: Kevles 1985. Kevles's book remains the authoritative history of eugenics nearly forty years after its publication, setting the stage for much of the more fine-grained work that followed it.

56 The branding and focus just shifted: Cowan 2008; Stern 2012; Comfort 2012.

57 Armed with maps of known genes' locations: Bates 2005; Tsui and Dorfman 2013.

57 As a result, linkage analyses: Lipner and Greenberg 2018.

58 Risch and Merikangas proposed seeking out: Risch and Merikangas 1996; Risch 2005.

58 "genome-wide association studies": Visscher, Brown, McCarthy, et al. 2012. For an excellent philosophical examination of the rise of genome-wide association studies in the study of behavior, see Schaffner 2016.

59 Peter Scheidt: Scheidt interviews, Sept. 25, 2015, April 6–7, 2017; Mark Klebanoff interview, Jan. 5, 2016.

61 The plan for what data: Landrigan, Trasande, Thorpe, et al. 2006.

61 The source of the most internal strife: Mark Klebanoff interviews, Jan. 5 and 8, 2016; Nigel Paneth interviews, Nov. 3 and Dec. 8, 2015.

62 "You have to drop our topic": Robert Michael interview, Feb. 19, 2016.

62 A series of contentious: The question of how to sample the children in the National Children's Study was discussed on a number of occasions between 2002 and 2004. Reports and meetings devoted to the topic include Westat (2002), "Sampling Strategies for the Proposed National Children's Study"; Battelle (2004), "Summary of Additional Probability-Based Studies Examined for Initial Response and Retention Rates"; "Final Report from the National Children's Study Sampling Design Workshop, March 21–22, 2004," May 9, 2004; Minutes of the National Children's Study Federal Advisory Committee (NCSAC) 10th Meeting, June 28–29, 2004; Interagency Coordinating Committee meeting minutes, June 29, 2004. One of the strongest voices in favor of the probability sample was the social scientist Robert Michael: R. T. Michael and O'Muircheartaigh 2008. Nigel Paneth and Mark Klebanoff were both forceful critics of the household-recruited probability sample: Nigel Paneth (2001), "Cohort/Longitudinal Studies in Children: Taking a Public Health Perspective" (presentation at Study Assembly of National Children's Study, Oct. 2001); Mark Klebanoff (2004), "The Epidemiologists Perspective" (presentation at National Children's Study Sampling Design Workshop, March 21–22, 2004).

63 The study designers believed a sample: Ogilvy Public Relations Worldwide 2007.

63 Alan Guttmacher: Guttmacher interview, Feb. 9, 2018.

64 Guttmacher had to look up information: Guttmacher interview, Feb. 28, 2022.

64 The result looked very little: "American Gene-Environment Study (AGES)," 9/16/2004 AGES Core Meeting: AGES Draft Proposal, Box 4 of 4 Deputy Director Files, Folder 18, NHGRI History of Genomics Program Archive, Bethesda, MD.

65 The sample sizes of the two studies: "The American Gene-Environment Study (AGES) or the United States Assessment of Heredity, Environment, and Lifestyle for Total Health (USA HEALTH)," 11/8/2004 EZ Potential Options—Large U.S. Cohort Study: AGES or USA HEALTH PowerPoint, Box 4 of 4 Deputy Director Files, Folder 24, NHGRI History of Genomics Program Archive, Bethesda, MD.

66 Collins asked to chat with Alexander: Collins email (Feb. 13, 2004), AGES Cohort Study 2/04–4/04: Emails about revised planning note for EZ; conversation with DA, Box 4 of 4 Deputy Director Files, Folder 33, NHGRI History of Genomics Program Archive, Bethesda, MD.

66 Alexander informed Scheidt: Sarah Keim and Mark Klebanoff interviews, June 12 and 13, 2017.

67 On March 2, Collins met Alexander: Collins emails (Feb. 13, 2004, and March 4, 2004), AGES Cohort Study 2/04–4/04: Emails about revised planning note for EZ; conversation with DA, Box 4 of 4 Deputy Director Files, Folder 33, NHGRI History of Genomics Program Archive, Bethesda, MD; Collins notes (March 2, 2004), AGES Cohort Study 2/04–4/04: Handwritten notes (March 2, 2004), Box 4 of 4 Deputy Director Files, Folder 33, NHGRI History of Genomics Program Archive, Bethesda, MD.

67 Collins next focused on packaging: Collins notes, AGES Cohort Study 2/04–4/04: Handwritten notes (March 17, 2004), Box 4 of 4 Deputy Director Files, Folder 33, NHGRI History of Genomics Program Archive, Bethesda, MD.

68 decades of environmental research: Oreskes and Conway 2010; Markowitz and Rosner 2013.

68 humans have only about 20,000–25,000 genes: Wade 2004. Sarah S. Richardson and Hallam Stevens provide a helpful introduction to this post–Human Genome Project realization in their edited volume, Richardson and Stevens 2015; see, in particular, their own chapter, "Beyond the Genome."

CHAPTER 3 INDUSTRY RELATIONSHIPS

70 Smeltertown first took shape: The most complete history of Smeltertown can be found in Perales 2010. Perales also published an article-length version of the story earlier in her Perales 2008. Elaine Hampton and Cynthia C. Ontiveros provide another detailed account of the events that led up to Smeltertown's demise in Hampton and Ontiveros 2019. A shorter version geared toward a nonacademic audience can be found in Villagran 2016.

70 Robert Safford Towne: Marcosson 1949.

71 Humans have been smelting for millennia: Morin 2010.

71 there was an obvious power imbalance: Perales 2010, chap. 2.

73 They saw the yellow clouds: Sullivan 2014.

73 El Paso city health officials: *Morbidity and Mortality Weekly Report* 1973.

73 the CDC agreed to send: The CDC didn't change its name to the Centers for Disease Control and Prevention until 1992.

73 Epidemic Intelligence Service teams: Thacker, Dannenberg, and Hamilton 2001.

73 to put Philip Landrigan in charge: Landrigan interview, Jan. 12, 2022.

74 The results, when they took the samples: Landrigan, Gehlbach, Rosenblum, et al. 1975.

74 Asarco was not sitting idle: Sullivan 2014, chap. 3.

74 He boarded a flight from El Paso: Landrigan interview, Jan. 12, 2022.

75 Children with elevated lead levels: Landrigan, Baloh, Barthel, et al. 1975.

75 Lawsuits soon followed: Sullivan 2014, chaps. 4 and 5.

75 Sadly, this story did not have a happy ending: Perales 2008 and 2010; Romero 1984. *The New York Times* covered the episode when it occurred in 1973: *New York Times* 1973.

75 Elevated lead levels continued: Díaz-Barriga, Batres, Calderón, et al. 1997.

76 Landrigan's experience with Asarco: D. Rosner and Markowitz 2005; Conway and Oreskes 2010; D. Michael 2008.

76 Scientific research in the United States: Bloom and Randolph 1990; Moses, Dorsey, Matheson, et al. 2005; Connelly and Propst 2006; Connelly and Propst 2007; Dorsey, de Roulet, Thompson, et al. 2010; Research!America 2017, 2018, 2019.

77 a concerted effort to commercialize science: Mirowski and Sent 2002; Geiger 2004; Washburn 2005; Mirowski 2011; Berman 2012.

78 *Pesticides in the Diets of Infants and Children:* Committee on Pesticides in the Diets of Infants and Children 1993.

78 Landrigan and Needleman together: Browner interview, July 2, 2021; Landrigan interview, Dec. 13, 2016; Ogilvy Public Relations Worldwide 2004, 32–33.

78 In 1849, they incorporated: Rodengen 1999; Lombardino 2000.

78 Pfizer expanded to become a leader: BBC 2014.

78 He was dying of cancer: Rodengen 1999, 80.

79 built its fortune on blockbuster drugs: Li 2014; Van Arnum 2012.

79 The strongest point in favor: Rebecca Henderson investigated the shift from blockbuster drugs to pharmacogenomics at Eli Lilly; see, for example, Henderson 2007 and 2008. A more industry-wide reflection on this evolution is available in Dugger, Platt, and Goldstein 2018. These paragraphs on the arguments for and against moving toward pharmacogenomics at Pfizer are based on interviews with Martin Mackay, June 15, 2020, and John Thompson, June 5, 2020, who were at Pfizer during that time. For more from the period itself, see Anand 2001.

81 In 1998 and 1999, it helped create: A. L. Holden 2002.

81 it formed business agreements: Webster, Martin, Lewis, et al. 2004; Hopkins, Ibarreta, Gaisser, et al. 2006.

81 the smaller company provided Pfizer: Business Wire 2004.

82 Collins's proposal to them: Pfizer memo, "Collins/Perlegen/Pfizer meeting," March 14, 2005; "An Unprecedented Partnership Opportunity to Unravel the Genetics of 20 Major Diseases."

82 Michael Leavitt: Leavitt interview, Nov. 9, 2015.

82 Collins began communicating: Collins email, March 6, 2005, Utah Genetics: Emails from FC about Utah, Box 4 of 4 Deputy Director Files, Folder 12, NHGRI History of Genomics Program Archive, Bethesda, MD; Collins notes from Mark Leppert conversation, March 9, 2005,Utah Genetics: Handwritten notes (March 9, 2005), Box 4 of 4 Deputy Director Files, Folder 12, NHGRI History of Genomics Program Archive, Bethesda, MD; Collins notes from Steve Prescott conversation, March 15, 2005,Utah Genetics: Handwritten notes (March 15, 2005), Box 4 of 4 Deputy Director Files, Folder 12, NHGRI History of Genomics Program Archive, Bethesda, MD; Prescott interview, June 23, 2015.

83 "Leavitt-friendly proposal": Collins email, March 17, 2005,Utah Genetics: Emails from FC with news, Box 4 of 4 Deputy Director Files, Folder 12, NHGRI History of Genomics Program Archive, Bethesda, MD.

83 Pfizer and Perlegen scientists traveled to NIH: "Unraveling the Genetics of Complex Disease: A Possible Resource for Case-Control Studies," June 30, 2005, Box 030, Folder 28, NHGRI History of Genomics Program Archive, Bethesda, MD; Pfizer memo, "NIH Proposal for Whole Genome Analysis on Major Diseases," July 10, 2005.

83 David Schwartz: Schwartz interview, March 30, 2016.

84 Zerhouni, Collins, and Schwartz's proposal: Elias Zerhouni, Francis S. Collins, and David Schwartz, "Genes, Environment, Health, and Disease: A Major New U.S. Initiative," Box 030, Folder 24, NHGRI History of Genomics Program Archive, Bethesda, MD.

84 The NIH delegation, as a result: Guttmacher email, Aug. 5, 2005; Scheidt notes, Aug. 22, 2005.

85 To his surprise and disappointment: Duane Alexander's address to the Interagency Coordinating Committee, Aug. 20, 2009.

86 In the meantime, there were two solutions: Scheidt interview, Oct. 13, 2016; Keim interview, Oct. 12, 2016.

87 brainstorming ways to share resources: Ogilvy 2004, 22; "A Shared Model of the National Children's Study and the American Genes and Environment Study," Aug. 18, 2005.

87 When Alexander met with the secretary: "Genes, Environment, Health, and Disease Across the Lifespan: A Major New U.S. Initiative," Meeting to Discuss Budget Issues, Department of Health and Human Services, Aug. 22, 2005.

87 the secretary made a special trip: *Deseret News* 2005.

87 Child Health announced the institutions: Initially, only six Vanguard Centers were announced. (1) The University of California at Irvine alongside Children's Hospital of Orange County was going to recruit the children in Orange County, California. (2) The Mount Sinai School of Medicine with the Mailman School of Public Health

at Columbia University was in charge of Queens, New York. (3) The rural county of Duplin, North Carolina, was going to be recruited by the team of Duke University, the University of North Carolina, and the Battelle Memorial Institute. (4) Three institutions in Philadelphia were given the opportunity to recruit participants from Montgomery County, Pennsylvania: Children's Hospital of Philadelphia, Drexel University School of Public Health, and the University of Pennsylvania. (5) The University of Utah was taking charge of Salt Lake County in Utah. (6) And Waukesha County, Wisconsin, was going to be reached by the University of Wisconsin, the Medical College of Wisconsin, Marquette University, Children's Service Society of Wisconsin, and the National Opinion Research Center. A seventh Vanguard site was announced soon after; a team from the University of South Dakota and the University of Cincinnati took charge of covering the remote regions of Brookings County, South Dakota, and the Lincoln, Pipestone, and Yellow Medicine Counties in Minnesota. An eighth Vanguard Center was initially planned for Orange County, Florida, but no site was ever selected. National Children's Study Steering Committee meeting, Nov. 29, 2005.

87 After the meeting Landrigan had dinner: Landrigan interview, Dec. 13, 2016; Clark interview, Dec. 2, 2015; Fleischman interview, Dec. 16, 2015.

88 The three agreed to stay in touch: Porter interview, Nov. 12, 2015.

89 the investment in gene hunting: Notes from the First Executive Committee Meeting of the Whole Genome Association Consortium, Nov. 29, 2005, Box 030, Folder 28, NHGRI History of Genomics Program Archive, Bethesda, MD; Pfizer memo, "Pfizer Support of the Whole Genome Association Consortium, a Public-Private Partnership Coordinated by the Foundation for the National Institutes of Health," Nov. 30, 2005.

89 No funds were included: Department of Health and Human Services, National Institutes of Health, National Institute of Child Health and Human Development, "FY2007 Budget," 35.

90 Fleischman stepped back from his desk: Fleischman interview, Dec. 16, 2015.

90 They agreed it was time: Landrigan interview, Dec. 13, 2016; Clark interview, Dec. 2, 2015.

90 The news that day involved: NIH 2006.

90 The reporters who packed into the Murrow Room: For video of the National Press Club event, see videocast.nih.gov.

91 When reporters were given the opportunity: The question session with reporters began around the thirty-eight-minute mark.

93 The months that followed were chaotic: Scheidt interview, Dec. 18, 2015.

94 Sarah Keim: Keim interview, Nov. 11, 2015. The "intern" designation became sullied by President Clinton's Monica Lewinsky affair. Lewinsky was not a participant in the Presidential Management Intern Program; she was an unpaid White House intern. But the terminological association was enough to rankle participants in the Intern Program. As a result, they changed it to the Presidential Management Fellows Program; on the name change, see Causey 1999.

94 "may react negatively": Keim email, Sept. 25, 2006.

95 But the shocking announcement: Duane Alexander's address to the Interagency Coordinating Committee, Aug. 20, 2009.

95 Fighting the prospect of termination: Clark interview, Jan. 7, 2022; Fleischman interview, Jan. 26, 2022.

95 He contacted current members: Porter interview, April 5, 2017.

96 Zerhouni faced the repercussions of Porter's advocacy: Elias Zerhouni testimony before House Subcommittee on Labor, Health, Human Services, and Education, April 6, 2006.

96 Zerhouni fared even worse: Elias Zerhouni testimony before Senate Subcommittee on Labor, Health, Human Services, and Education, May 2006. A transcript of the Senate event can be found here: www.govinfo.gov. For media coverage of the growing congressional support for the National Children's Study, see Commonwealth Fund 2006; "Matsui Urges Appropriators to Save the National Children's Study," 2006.

96 Analyses performed for the study leadership: Peter Scheidt, "National Children's Study: Update and Report to NCSAC," May 31, 2006.

96 An editorial in *Scientific American: Scientific American* 2006.

CHAPTER 4 FROM ARTIFICIAL NOSE TO GENOMICS JUGGERNAUT

98 Illumina is without peers: Many of the details regarding Illumina's size, scope, and value can be found in its annual reports, as well as documents like "Illumina: At a Glance," available on the Illumina website.

98 "the Google of genetic testing": Zhang 2016.

99 artificial nose business: M. Jones 2014. Anthony Czarnik, one of the Illumina cofounders, also recounts a version of this in Czarnik 2019.

99 He then took the device: Barnard and Walt 1991. The EPA's site on Pease is available here: cumulis.epa.gov. The Pease site for the CDC is at www.atsdr.cdc.gov. There are ongoing investigations into the health impacts of the people stationed at Pease when the contamination occurred; see, for example, Ropiek 2018; Leigh 2020.

100 The design of the first working electronic nose: Persaud and Dodd 1982.

100 It was an exercise in biomimetics: Walt and his graduate students Shannon E. Stitzel and Matthew J. Aernecke provide a useful introduction to the history and basic elements of an artificial nose in their D. R. Walt, Stitzel, and Aernecke 2012.

100 They fabricated a bundle: Dickinson, White, Kauer, et al. 1996.

101 described as a "bead array": K. L. Michael, Taylor, Schultz, et al. 1998. Deirdre Bradford Parsons tells the story of the microwell and then bead array discoveries in her Parsons 2007, chap. 4.

101 A fiber-optic bundle only one millimeter: Dickinson, Michael, Kauer, et al. 1999.

102 The bead array was a breakthrough: Walt referenced the genetic applications (and the vapor sensor applications) of the bead array at the very end of the first publication that announces it, K. L. Michael, Taylor, Schultz, et al. 1998, 1248, "We are presently extending this work to demonstrate arrays for immunoassays, gene probe sequences, and vapor sensors."

103 After just three days, Chee told Walt: Walt interview, Jan. 19, 2021.
103 set aside fears: Czarnik 2019.
103 needed to act fast on the genetics side: Walt interview, Jan. 19, 2021.
103 The Illumina scientists spent: Parsons details the rollout of Illumina in Parsons 2007, chap. 5. David Stipp gives a snapshot of the buzz surrounding Illumina in the months before it went public in Stipp 2000.
104 Illumina, as a result, positioned itself: Dickinson interview, Jan. 20, 2021.
104 Illumina acquired Spyder Instruments: Lebl 2003.
105 Illumina acquired Solexa: GenomeWeb News 2007.
105 International HapMap Project: Thorisson, Smith, Krishnan, et al. 2005; McVean, Spencer, and Chaix 2005; Manolio, Brooks, and Collins 2008. On the various centers' reliance on Illumina machines, see Parsons 2007, 265–66.
105 Electronic Medical Records and Genomics (eMERGE) Network: McCarty, Chisholm, Chute, et al. 2011.
106 23andMe, backed by Google's deep pockets: Goetz 2007. Sean Valles 2012 was an early advocate of subjecting direct-to-consumer genetic testing to government regulation.
106 Illumina soon formed partnerships: Dolan 2009.
106 Indeed, it soon became difficult: Manolio and Collins 2009; Corvin, Craddock, and Sullivan 2010.
107 The genetic effects for the vast majority: Maher 2008; Turkheimer 2011; Matthews and Turkheimer 2021; Richardson and Stevens 2015.
107 The solution, they countered: Manolio, Collins, Cox, et al. 2008; Schork, Murray, Frazer, et al. 2009.
109 After the meeting, the group gathered: Roderick recalls this history in Kuska 1998.
109 get in on the big-data revolution: Christopher Phillips tells the story of the statistical, epidemiological, and computing methodologies that set the stage for the rise of big data in his forthcoming book, *Number Doctors: The Emergence of Biostatistics and the Reformation of Modern Medicine.*
110 Even environmental health scientists: D. R. Walt 2005; Patel, Bhattacharya, and Butte 2010.
110 to get the NIH in the exposome business: See the NIEHS's site on "Exposure Biology and the Exposome" at www.niehs.nih.gov.
110 Schwartz and Walt even had conversations: Walt interview, Oct. 28, 2020.
111 initially took the optical nose: Albert interview, Jan. 8, 2021; Stitzel interview, Jan. 11, 2021.
111 detect nitroaromatic compound vapors: Albert and Walt 2000.
111 distinguish French roast coffee from Colombian: Albert, Walt, Gill, et al. 2001.
111 a massive model of a dog's nose: Stitzel, Stein, and Walt 2003; Pearson 2003.
111 The future of the optical nose: Anthes 2008.
111 what they called their "portable sniffer": Albert, Myrick, Brown, et al. 2001.
111 The task for Albert was to see: Albert interview, Jan. 8, 2021; Bochnak 2002.
113 Portability was just one of the challenges: Albert interview, Jan. 8, 2021; Stitzel interview, Jan. 11, 2021; D. R. Walt 2005.
113 There were financial ones, too: Walt interviews, Oct. 28, 2020, and Jan. 19, 2021.

Walt's lab was by no means the only artificial nose group struggling to find financial investors; Emily Anthes 2008 quotes another researcher as saying that the science was "just waiting for somebody with deep pockets to come up with the money and focused effort to develop these noses commercially."

114 Walt, facing this practical and financial reality: A handful of publications regarding the optical nose continued to trickle out of the Walt lab after 2008 (see, for example, D. R. Walt, Stitzel, and Aernecke 2012), but those were reviews of research rather than reports on new efforts under way.

114 Illumina saw the same dynamic: Dickinson interview, Jan. 20, 2021.

114 Fitbit, for example, went live in 2007: Bora 2018; C. Marshall and Stables 2020.

115 It was a terrifying public health threat: *Morbidity and Mortality Weekly Report* 1996.

115 History and literature are replete: Cowgill 2018; Klass 2020, chap. 10; Task Force on Sudden Infant Death Syndrome 2005.

116 Harold Abramson: Abramson 1944.

116 British pediatrician David Davies connected: D. P. Davies 1984.

116 Dutch researchers went back and scrutinized: Engleberts and de Jonge 1990.

117 Those observational studies were enough: Fleming, Gilbert, Azaz, et al. 1990; Dwyer, Ponsonby, Newman, et al. 1991. Ruth Gilbert, Georgia Salanti, Melissa Harden, and Sarah See 2005 document how there was clear epidemiological evidence suggesting that prone sleeping increased risk of SIDS by 1970.

117 Back to Sleep campaign: *New York Times* 1994.

117 When a caller phoned in: "Child Health" 2002.

117 The results of the campaign: Willinger, Hoffman, Wu, et al. 1998.

118 SIDS deaths, in turn, plummeted: *New York Times* 1996. In 2012, NICHD replaced the Back to Sleep campaign with the Safe to Sleep public education campaign: safetosleep.nichd.nih.gov.

118 Just one decade after the Back to Sleep campaign: Decades of data on SIDS in the United States, revealing the steep decline that started in the early 1990s, can be found on the CDC's website: www.cdc.gov.

119 The dozen experts in public health: Samuel Preston interview, July 13, 2016; Connie Citro interview, Aug. 11, 2016.

119 The report that the group released: Panel to Review the National Children's Study Research Plan 2008, 131.

120 The response from Child Health: "National Children's Study Response to the National Academy of Sciences Review of the National Children's Study Research Plan."

120 In Utah, Edward Clark was meeting: Clark interview, Jan. 14, 2016.

121 "Unfortunately, we are all sticking": Keim email, May 17, 2007.

CHAPTER 5 THE POLITICS OF THE PERSONAL

123 "note on the desk": U.S. Department of Health and Human Services 2008, 17–18. For media coverage of the report and Leavitt's characterization of it as a "note

on the desk" to the next administration, see Lauerman 2008 and L. Taylor 2008. Leavitt's "note on the desk" reference was a nod to the common practice of presidents leaving a letter for their successor in the Oval Office.

123 Richard Nixon created the U.S. Environmental Protection Agency: Flippen 2000.

124 the dynamic had radically changed: Andrews 2006; Layzer 2012; Turner and Isenberg 2018.

124 Nixon feared a Democratic challenger: Flippen 2000.

125 After Nixon was reelected: Layzer 2012, chap. 3.

125 Nixon, for example, was furious: Flippen 2000, 172.

125 The ban proceeded, but Ruckelshaus: Ruckelshaus, after leaving the EPA, made another famous appearance toward the very end of the Nixon administration. After serving in the FBI, he was moved to deputy attorney general, where he was one of the people, on October 20, 1973, who refused Nixon's demand to fire the Watergate special prosecutor and resigned instead. The "Saturday Night Massacre" was soon followed by the initiation of impeachment proceedings in the House of Representatives.

125 influential conservative think tanks: Andrew 2004; Schulman and Zelizer 2008; Jacques, Dunlap, and Freeman 2008; Stahl 2016.

126 Reagan's administration used executive actions: Lash, Gillman, and Sheridan 1984; Layzer 2012, chap. 4; Andrews 2006, chaps. 13 and 14; Turner and Isenberg 2018, chaps. 2 and 3.

126 The humiliating rebuke: The resulting public and congressional scrutiny is recounted in all the works that tell this history. Examples from the era in the press can be found in Shabecoff 1982; Romano 1983; Romano and Trescott 1983.

127 By the time Bill Clinton left office: Andrews 2010; Andrews 2006, chap. 16; Layzer 2012, chap. 6.

127 When Bush took the White House: Devine 2004; Fredrickson, Sellers, Dillon, et al. 2018; Andrews 2006, chap. 16; Layzer 2012, chap. 7.

128 Carol Browner: Browner interview, July 2, 2021.

128 President Clinton's 1994 executive order: Executive Order 12898.

128 contributed to an executive order from Clinton: Executive Order 13045.

128 it was during a 1998 gathering: Marshalyn Yeargin-Allsopp interview, Feb. 1, 2016; Mark Klebanoff interview, Jan. 5, 2016; Klebanoff 2009.

128 That proposal, in turn: Children's Health Act of 2000; Clinton 2000.

129 Zerhouni, arguing as late as September 2008: *Risk Policy Report* 2008.

129 Zerhouni stepped down as NIH director: Wadman 2008.

129 Alexander went to the NIH's budget office: Duane Alexander's address to the Interagency Coordinating Committee, Aug. 20, 2009.

129 put together a menu of proposals: Keim interview, Nov. 23, 2015.

130 the very same features: The most in-depth looks at this neoliberal appeal of genomics can be found in Rajan 2006 and Dickenson 2013. But see also Joseph and Mrdjenovich 2017.

130 straightforward embrace of individualism: Dickenson 2013 spends a great deal of time probing the way that personalized medicine packages a neoliberal-friendly form of narcissism.

130 A report released by Health and Human Services: U.S. Department of Health and Human Services 2007, 2008. Secretary Leavitt provided a foreword to the 2007 report and a prologue to the 2008 report that give some insight into his personal views on the topic. See also Leavitt and Downing 2008. Leavitt's Secretary's Advisory Committee on Genetics, Health, and Society also issued a 2008 report of its own, which was very much an exercise in promoting personalized medicine.

131 personalized medicine gave the illusion: See Vogt, Green, and Brodersen 2018 for a critical take on this notion that patients are somehow empowered by personalized medicine. To be true, the physician-patient relationship may indeed be impacted by the spread of personalized medicine, but the impact will be less about more empowered patients and more about greater uncertainty in diagnoses, a loss of privacy, and greater ambiguity between what it means to be a patient and what it means to be a research participant; on these features, see Eyal, Sabatello, Tabb, et al. 2019.

131 The President's Council: President's Council of Advisors on Science and Technology 2008.

131 And the Republican Party platform: Republican Party Platform 2008, 39.

131 According to Obama's health policy adviser: Hughes interview, Dec. 2, 2016; "Interview with Senator Barack Obama About Genomics and Personalized Medicine Act of 2006" 2007.

132 Senator Obama introduced: Senator Obama introduced the Genomics and Personalized Medicine Act of 2006 in the 109th Congress (S.3822). Media reports on the bill can be found in Technology Networks 2006 and Swain 2006. Senator Obama reintroduced the bill with the co-sponsor Richard Burr (R-N.C.) in the 110th Congress of 2007 (S.976). For an analysis of the differences between the 2006 and the 2007 bills, see Lee and Mudaliar 2009.

132 advocates of personalized medicine: The Obama-Biden Plan to Combat Cancer includes a section titled "Support Advances in Personalized Medicine," which highlights his sponsorship of the Genomics and Personalized Medicine Act and then adds, "As president, Obama will continue to support advances in personalized medicine to help ensure early detection and treatment of cancer and other diseases." GenomeWeb 2008; Abrahams 2008.

132 "personalized medicine president": R. C. Wells 2009.

132 His open embrace of religion: Collins 2006.

132 Collins pressed hard to persuade Congress: Collins was by no means the only voice pushing for a genetic nondiscrimination bill, but he and the National Human Genome Research Institute were powerful and public advocates; Collins, in fact, called for legislation at the April 2003 Human Genome Project celebration at the Smithsonian.

133 Congress ultimately passed: For scholarly reflections on the Genetic Information Nondiscrimination Act, see J. L. Roberts 2010 and 2011; Rothstein 2008; Allison 2008; E. A. Feldman 2011.

133 In August 2008, Collins stepped down: Greenemeier 2008.

133 Collins joined the Obama-Biden transition team: Burns 2008.

134 looking shell-shocked: Keim interview, Nov. 23, 2015.

134 "deception or incompetence": Scheidt notes, Jan. 14, 2009; Scheidt interview, March 8, 2016.

134 The investigation, Kington told Alexander and Scheidt: Kington memo to Alexander, "The National Children's Study, Eunice Kennedy Shriver National Institute of Child Health and Human Development," March 11, 2009.

134 On the evening of Monday, March 16: Scheidt notes, March 16, 18, and 19, 2009; Scheidt interviews, Oct. 13, 2016, April 7, 2017.

135 NIH personnel, ten days later: Rep. David R. Obey Holds a Hearing on National Institutes of Health: Budget, Implementation of Recovery Act, and National Children's Study, March 26, 2009, House Appropriations Committee, Subcommittee on Labor, Health and Human Services, Education, and Related Agencies. Video of the hearing is available here: www.youtube.com.

137 After the hearing, the science reporter: Wadman 2009.

137 A recruiter in South Dakota: Bonny Specker interview, Nov. 17, 2016.

137 The directors reported a variety: Barbara Entwisle interview, Nov. 18, 2016; Landrigan interview, Dec. 13, 2016.

138 President Obama, on July 8: White House 2009. Rumors circulated about a Collins nomination for several months in advance; see, for example, C. Holden 2009 and G. Harris 2009.

138 the Office of Management Assessment released: Office of Management Assessment 2009. The report was obtained by way of a Freedom of Information Act request filed with the National Institutes of Health in 2016.

139 Rather, the investigators pointed: Office of Management Assessment 2009, 14; Keim email, Sept. 25, 2006; Keim email, May 17, 2007. A source familiar with the investigation who asked not to be identified pointed to the emails in particular as the damning pieces of evidence. Regarding documentation of the budget directive, the program office staff at Child Health did find a 2006 email from NIH budget officers telling them not to change the cost estimate for the study; however, the Office of Management Assessment report judged that directive to be about cost reporting only in that one budget cycle, not anything more general than that.

139 Alexander hand delivered the letter: Scheidt interview, Oct. 13, 2016.

139 "More Turmoil over Hidden Costs": Kaiser 2009a. The accusations of deliberate deception particularly irked those in the National Children's Study community who knew Alexander and Scheidt well. As one associate put it, "Those two don't have a deceptive bone in their bodies. They wouldn't be able to lie to their wives about having a girlfriend."

140 "This is a turbulent time": Duane Alexander's address to the Interagency Coordinating Committee, Aug. 20, 2009.

143 Alexander mustered his pleasant demeanor: An NIH videocast of the September 21 advisory council meeting where Alexander made the announcements is available here: videocast.nih.gov.

144 He was going to be stepping down: Alexander's announcement regarding his own

departure begins about fifteen minutes into the video recording. Neither Scheidt nor Alexander left federal employment immediately. Scheidt stepped down from the National Children's Study to serve as a special adviser to the Child Health director for a short time before returning to Children's National Medical Center in Washington, D.C., where he worked before signing on to direct the National Children's Study in 2000. Alexander moved over to the Fogarty Center, a center at the NIH devoted to global health research, where he briefly acted as an adviser there on child and maternal health, before retiring.

145 That same fall, Jocelyn Kaiser: Kaiser 2009b.

CHAPTER 6 GENOMICS FINDS ITS HAMMER

146 Congress demanding the review: Duncan, Kirkendall, and Citro 2014, 1.

146 "law of the instrument": Kaplan first introduced the law of the instrument at the American Educational Research Association conference banquet in 1962 in a presentation titled "The Methodology of Social Research"; Milton Horowitz recounts that "highlight of the 3-day meeting" in Horowitz 1962. Kaplan worked the idea into his own publications by 1964 in A. Kaplan 1964a and 1964b.

146 the cover of *Time* magazine: The May 6, 1966, issue of *Time* bears Kaplan's image on the cover (alongside nine other "Great Teachers"), and the discussion of him occurs on page 85.

146 documentaries about inspirational educators: The National Educational Television and Radio Center produced the documentary *Men Who Teach 3: Abraham Kaplan* in 1968.

146 "Give a boy a hammer": Abraham Maslow 1966 made a similar observation about hammers and nails; as a result, the law of the instrument is sometimes also referred to as Maslow's hammer.

147 a lock of hair from a man: M. Rasmussen, Li, Lindgreen, et al. 2010.

148 fingertip bone found in Siberia's Denisova Cave: Krause, Fu, Good, et al. 2010.

148 draft sequence of the Neanderthal genome: R. E. Green, Krause, Briggs, et al. 2010.

148 snippets of mitochondrial DNA: Cann, Stoneking, and Wilson 1987.

148 These new studies: Svante Pääbo, lead author on both the Denisova publication and the Neanderthal draft genome publication, was awarded the 2022 Nobel Prize in Physiology or Medicine for his contributions to the research that laid the groundwork for paleogenomics.

148 "rewrite human history": Dalton 2010.

148 "Good Year for Neanderthals (and DNA)": Joyce 2010.

148 Stories of newly discovered humans: Elizabeth D. Jones and Elsbeeth Bösl 2021 provide an analysis of how media hype surrounding the ancient DNA excitement compared with earlier media attention of genetic work pertaining to human history. E. Jones 2019 also reflects on this version of "celebrity-science" that took shape around aDNA research.

149 Reich contributed: Reich, Green, Kircher, et al. 2010; Meyer, Kircher, Gansauge,

et al. 2012; Prüfer, Racimo, Patterson, et al. 2014; Haak, Lazaridis, Patterson, et al. 2015.

149 "gold rush": *Edge* 2016. Reich eventually pulled together much of his contributions to the aDNA research alongside his reflections on its revolutionary nature in Reich 2018. On the claims of "revolution" surrounding aDNA research, see Källén, Mulcare, Nyblom, et al. 2021.

149 offered up sweeping theories: Ewen Callaway 2018 provides a very insightful review of this tense dynamic that played out in archaeology upon the arrival of ancient DNA research by geneticists. Another entry point into the way that Reich and the other geneticists upended archaeology can be found in the excellent article that Gideon Lewis-Kraus 2019 wrote for *The New York Times.*

149 "We haven't gone through graduate school": *Edge* 2016.

150 Gustaf Kossinna embodied those ideas: Baudou 2005; Hakenbeck 2008; Maner 2018.

150 Nazi leaders eagerly embraced: Arnold 1990 and 2006; Mees 2008.

150 Starting in the 1960s, archaeologists: Heyd 2017; Furholt 2018; Veeramah 2018.

151 The theoretical reorientation within archaeology: Trigger 1989; A. L. Johnson 2004; Renfrew and Bahn 2005.

152 they focused upon those topics: Downes 2019.

152 genetic information about different individuals: Monika Piotrowska 2009 provides a very useful assessment of what these comparative genomics claims mean.

152 the Saqqaq: Much of what we now know about the ancient Saqqaq people and their culture comes from research led by Bjarne Grønnow, who first discovered the artifacts that led to the archaeological work conducted at Qeqertasussuk: Grønnow 1988, 2012, and 2016. See also Meldgaard 2004.

152 limits of research on ancient DNA: Gibbons 2011 and 2015; Harmon 2012; Rogers, Bohlender, and Huff 2017.

153 a patrilocal residential arrangement: Knipper, Mittnik, Massy, et al. 2017. For endorsements of this interdisciplinary style of work between geneticists and archaeologists, see Veeramah 2018, Sawchuk and Prendergast 2019, and Downes 2019.

154 Archaeologists were critical: Horsburgh 2015; Heyd 2017; Furholt 2018; Veeramah 2018; Tsosie, Begay, Fox, et al. 2020; Hawks 2021; Cortez, Bolnick, Nichols, et al. 2021.

154 This genomic imperialism: Mäki, Walsh, and Pinto 2018 provide a variety of historical and philosophical perspectives on how scientific imperialism works.

156 *G Is for Genes:* Asbury and Plomin 2013.

156 the field of behavioral genetics: Histories of the science of behavioral genetics can be found in Panofsky 2014, Tabery 2014, and Schaffner 2016.

156 Initial hope in the 1990s: Plomin and Neiderhiser 1991; Plomin, Kennedy, and Craig 2006.

156 turning to genome-wide association studies: Davis, Butcher, Docherty, et al. 2010.

156 In 2013, a large, multinational research team: Rietveld, Medland, Derringer, et al. 2013. See pages 97–108 of the supplement to this publication for the list of cohorts

included and their heavy reliance on the Illumina platforms. This paper was just the first in a series of genome-wide association studies performed by the Social Science Genetic Association Consortium—deemed "EA1" for their first study of educational attainment; it was then followed up by EA2 in 2016, which included nearly 300,000 participants, and EA3 in 2018 involving over 1 million individuals; EA4 is forthcoming. Each iteration identified more and more SNPs but still accounted for only a fraction of the variation in educational attainment (or cognitive performance). Eric Turkheimer, another behavior geneticist, has argued more forcefully and sustainably than anyone that these gene-hunting exercises by social scientists are destined to fail when it comes to identifying actual genes that can be sources of reliable interventions: Turkheimer 1998, 2000, 2011, 2014, 2015, 2016, 2019a; Matthews and Turkheimer 2021.

157 Personalized nutrition: Reinagel 2012.

157 personalized skin-care products: Younom (www.younom.com) offered skin-care products based on genetic information about customers.

157 personalized matchmaking: DNA-guided dating sites from the era included Instant Chemistry (www.instantchemistry.com), DNA Romance (www.dnaromance.com), and SingldOut (www.singldout.com).

157 *G Is for Genes* was just the first book: Wilby 2014; Gaysina 2016; Galeon 2016; Hart 2016; Kovas, Malykh, and Gaysina 2016; Bouregy, Grigorenko, and Latham, 2017. A noteworthy precursor to the explosion in this literature in 2013 is Grigorenko 2007. The EA1/EA2/EA3 authors, for their part, made it a point to say that their results were not a route to personalized education for individual children. For a different approach to incorporating genetic information into the educational sphere that doesn't take the form of "personalized education," see Harden 2021.

157 Plomin took aim: Plomin found himself in hot water around the same time that *G Is for Genes* was published when it emerged that he was teaming up with scientists at BGI in China to deliver them samples from brilliant participants in the Study of Mathematically Precocious Youth so that BGI could scan their genomes (with Illumina machines) in search of genes associated with intelligence; for that story, see Yong 2013, Naik 2013, and Specter 2014.

158 Critics throughout the 1970s and 1980s: Rosenbaum 1976; Oakes 1982 and 1985.

158 These practices didn't just mirror: Rist 1970; Wheelock 1992; Hanushek and Wößmann 2006.

159 That trend, however, gradually subsided: Tom Loveless 1999 wrote about the controversy surrounding tracking; he then led a Brookings Institution report on the revival of ability grouping in the first decade of the twenty-first century with Loveless 2013.

159 Teachers didn't lack awareness: Panofsky 2015. I'm also grateful to Dawn-Marie Tabery, a public school sixth-grade teacher, for pointing out how, from a teacher's perspective, the problem isn't a lack of awareness about the virtues of personalizing education; rather, it's a lack of resources to do it effectively for all students.

160 educational reform guided by DNA: The educational scholar Daphne Martschenko, the bioethicist Maya Sabatello, and the philosopher Lucas Matthews make these

points in a series of important articles: Martschenko, Trejo, and Dominque 2019; Martschenko 2020a and 2020b; Martschenko and Matthews 2020; Matthews, Lebowitz, Ottman, et al. 2021; Sabatello 2018; Sabatello, Insel, Corbeil, et al. 2021. See also D. Roberts 2015, Panofsky 2015, and Turkheimer 2019b.

161 "You may think of me": Collins made a public reference to this "Genome Guy" line in his first town hall with NIH employees on August 17, 2009; a videocast of the event is available here: videocast.nih.gov.

161 made them priorities for the whole agency: Collins 2010a and 2010b.

161 Media interest in it: Evans 2010.

161 *The Language of Life:* Collins 2010c.

161 Personalized medicine, Collins claimed: Kaiser Health News 2010.

161 replacing the one-pill-fits-all approach: Collins 2010d.

162 "national highway system": Hamburg and Collins 2010. Collins welcomed attendees at a May 2010 meeting at the NIH called "Health Economics: NIH Research Priorities for Health Care Reform," and he referenced the role that large, longitudinal cohort studies like the National Children's Study could play in that effort as part of his opening remarks; the meeting summary is available online at commonfund.nih.gov.

162 Alan Guttmacher: The National Children's Study Advisory Committee Meeting minutes, Oct. 14, 2010.

163 Throughout 2011, though, those projects: "National Children's Study Formative Research Projects That Have OMB Clearance," May 6, 2011; "The National Children's Study Research Day: Come Learn, Collaborate, and Innovate with Us!," July 7, 2011.

164 sample size for the National Children's Study: "DRAFT Concept of National Children's Study Main Study," July 20, 2011; Steven Hirschfeld, "National Children's Study: Briefing for NCS Federal Advisory Committee," July 20, 2011; Ruth Brenner, "Draft Concept of the National Children's Study Main Study," July 20, 2011.

164 Ellen Silbergeld and Jonas Ellenberg: National Children's Study Federal Advisory Committee Meeting minutes, July 20, 2011.

165 The working groups no longer met: Hirschfeld announced the news regarding the working groups in Steven Hirschfeld, "National Children's Study Interim Summary and Plans," Nov. 1, 2009.

165 The study leadership at Child Health: Sarah Keim and Mark Klebanoff, members of the Interagency Coordinating Committee at that time, recalled how Hirschfeld stopped attending those meetings (Keim and Klebanoff interview, June 12, 2017).

165 The site directors now wrote: Nigel Paneth interviews, Oct. 20 and Dec. 8, 2015, Feb. 18, 2016; Jeff Murray interviews, Feb. 3 and March 2, 2016; Stephen Buka interviews, Nov. 28 and Dec. 20, 2016; "The Scientific Direction of the National Children's Study," a letter from Michael Bracken, Stephen Buka, Irva Hertz-Picciotto, George Lister, and Nigel Paneth to Alan Guttmacher and Steven Hirschfeld, Sept. 21, 2011.

165 the study would no longer utilize: Meredith Wadman first reported on the controversy surrounding the switch to a convenience sample in Wadman 2012a. The

details for the plan appeared in Office of the Director, National Institutes of Health, FY2013 Budget, 34–35.

165 Silbergeld publicly resigned: Silbergeld interview, Nov. 17, 2016; Wadman 2012b.

165 Ten days later, Ellenberg resigned: Kaiser 2012; Scudelleri 2012.

165 Sarah Keim and her husband: Keim and Klebanoff interview, June 12, 2017. Keim, Klebanoff, and the Vanguard directors were not alone. The social scientist Robert Michael, another longtime member of the Federal Advisory Committee, who was among the most influential voices in favor of the national probability sample, cut his ties to the study after Alexander and Scheidt were dispatched. He recalled, "I lost confidence in the study and had little respect for those who had dismissed them." Michael interview, Feb. 19, 2016. Donald Mattison, who helped craft the Children's Health Act and served as the first chair of the Federal Advisory Committee, also left in 2012; Mattison interview, Dec. 21, 2016.

166 Collins appeared before the Senate: U.S. Senate Subcommittee on Appropriations, Departments of Labor, Health, and Human Services, and Education, and Related Agencies for Fiscal Year 2013, Department of Health and Human Services, National Institutes of Health, March 28, 2012. Senator Shelby's questions begin on page 121 of the record.

166 The public scrutiny that kicked off: Wadman 2012c; Struck 2013.

167 they never dialed back the annual expenses: Duncan, Kirkendall, and Citro 2014, 16.

167 "night and day": Citro interview, Aug. 29, 2016.

167 The second time, the group received: Duncan, Kirkendall, and Citro 2014, 24–25.

168 the new leadership at NIH: Guttmacher, Hirschfeld, and Collins 2013.

168 one long indictment: Duncan, Kirkendall, and Citro 2014. These particular quotations are pulled from the summary, which can be found on pages 2–5.

169 In a highly unusual move: Ibid., app. and chap. 6.

169 "The study leadership were trying": Citro interview, Aug. 29, 2016.

169 Advisory Committee to the Director: Philip Pizzo interview, July 26, 2016; Russ Altman interview, July 6, 2016.

169 "feasible, as currently outlined": National Children's Study Working Group 2014, 3.

169 The experts that Collins gathered: Russ Altman interview, July 6, 2016; Philip Pizzo interview, July 26, 2016; Lynn Marie Miranda interview, Sept. 19, 2016.

169 "as currently outlined, is not feasible": National Children's Study Working Group 2014.

169 The NIH director said: Collins 2014b. Terminating the National Children's Study did not equate to abandoning environmental health research at the Child Health Institute. With the funds that were to be allocated to the National Children's Study, the NIH set up the Environmental influences on Child Health Outcomes (ECHO) program at Child Health, which served to pull together smaller pediatric research cohorts across the United States in an effort to combine those populations into one large research cohort. Still, ECHO was a significant departure from the National Children's Study. Lumping together a collection of preexisting cohorts (a "cohort of

cohorts" approach) creates all sorts of logistical, ethical, and scientific challenges. Different cohorts will have recruited their participants at different times and with different sampling strategies; they will have collected data in different ways; they will have monitored different health outcomes; the participants (or their parents) will have consented to different things. As a result, it can be quite difficult to integrate the studies in such a way that data from all the cohorts can be brought to bear on hypotheses of interest. The most straightforward way to make this point in the context of this book is to point to the fact that Collins, when designing the American Gene Environment Study, briefly considered going the cohort of cohorts route for his genomics study, only to quickly embrace the single, large, longitudinal study as far superior.

CHAPTER 7 DNA'S "DIRTY LITTLE SECRET"

171 MaryAnne DiCanto: I first encountered DiCanto's story in Szabo 2018. Szabo kindly reached out to DiCanto's husband, Scott Primiano, on my behalf, and he agreed to speak with me. The material in this chapter on DiCanto's experience with cancer is based on my conversation with Primiano on November 11, 2019, as well as email communications. DiCanto, however, did a number of interviews herself, and so elements of this story can also be found in Choudhury 2015, Sipher 2015, Millar 2015, Chan 2016, Ruiz 2016, and Rowley 2016.

172 METAvivor and the Tutu Project: METAvivor's website is www.metavivor.org. Information on the Tutu Project can be found at thetutuproject.com.

173 The branding was so successful: Joan Fujimura refers to this phenomenon as the "bandwagon effect"; her focus was the rise of genetic explanations in cancer research, but it is equally applicable to personalized medicine; see Fujimura 1988.

173 it quickly became adopted: An early example of the expansion of the personalized medicine concept beyond pharmacogenomics can be found in Bottles 2001.

174 Personalized medicine journals were created: *Personalized Medicine* was first published in 2004; in 2007, the *Mount Sinai Journal of Medicine* renamed itself the *Mount Sinai Journal of Medicine: A Journal of Translational and Personalized Medicine. Current Pharmacogenomics,* first published in 2003, renamed itself *Current Pharmacogenomics and Personalized Medicine* in 2008. And the similarly named *Pharmacogenomics and Personalized Medicine* also first appeared in 2008.

174 academic citations of the term: References to citations about personalized medicine are based on a PubMed search for ("personalized medicine" or "personalized medicines" or "personalised medicine" or "personalised medicines"). The year-by-year publication count for that first decade looks like this: 1999 (3), 2000 (9), 2001 (10), 2002 (31), 2003 (41), 2004 (72), 2005 (120), 2006 (132), 2007 (208), 2008 (296).

174 Personalized Medicine Coalition: Munroe 2004; Abrahams, Ginsburg, and Silver 2005; Abrahams 2005.

174 All that success brought scrutiny: Eric Juengst, Michelle L. McGowan, Jennifer R. Fishman, and Richard A. Settersten Jr. 2016 undertook an enormous study,

interviewing more than 140 proponents of genomic medicine to understand their satisfaction or dissatisfaction with the phrase "personalized medicine." They document the concerns about hype and the concerns about physicians always having personalized medicine, but it's really the focus on how "personalized medicine" misleads about the nature of how it works that raises the most worries. See also Solomon 2016 and Fuller 2017 for additional reflections on how personalized medicine rolled out, along with its relationship to earlier concepts like evidence-based medicine and narrative medicine.

174 "individualized drug therapies": Eichelbaum, Ingelman-Sundberg, and Evans 2006; Daly 2007.

174 "individual, genetics-based approach": Mancinelli, Cronin, and Sadée 2000. For other invocations of the "individualized" locution, see Xie and Frueh 2005, Shastry 2006, Waldman and Terzic 2008, and R. Roberts 2008.

174 "Personal Pills": Stix 1998.

174 "Just for You": K. F. Schmidt 1998.

175 It was the garment equivalent: The tailoring (dis)analogy has become quite common in discussions of personalized medicine. Adam Hedgecoe 2004 made very early reference to it, but see also Prainsack 2014.

175 The Hippocratic Corpus: Skiotis, Kalliolias, and Papavassiliou 2005; Konstantinidou, Karaglani, Panagopoulou, et al. 2017; Stegenga 2018.

175 "Give different ones to different patients": Hippocrates 1998, 157.

175 natural variation in shoulder sockets: Hippocrates 1968, 213–15.

175 individual differences in the course of pneumonia: Hippocrates 1988, 27.

175 the HLA match between donor and recipient: Thorsby 2009; Howard, Fernandez-Vina, Appelbaum, et al. 2015.

175 incorporation of blood type into transfusions: Starr 1998; Giangrande 2000.

176 the administration of general anesthesia: Snow 2008; Zambouri 2007.

176 Daniel Nebert, one of the pioneers: Nebert, Jorge-Nebert, and Vesell 2003; Nebert, Zhang, and Vesell 2008.

177 "What we know about the genome": Dreifus 2008.

177 The report that the group issued: Committee on a Framework for Development of a New Taxonomy of Disease 2011. The fact that the charge to the committee came from Collins is referenced on page 10 of the report.

177 The choice of "precision medicine": Maynard Olson interview, Oct. 11, 2019; Stephen Galli interview, Oct. 17, 2019. Maynard Olson tells elements of this story in Olson 2017.

177 In *Phaedrus,* Plato tells the story: Plato 1925, 265e. A similar story can be found in a Taoist allegory from Zhuang Zhou, where an experienced butcher explains, "A good cook need sharpen his blade but once a year. He cuts cleanly. An awkward cook sharpens his knife every month. He chops. I've used this knife for nineteen years, carving thousands of oxen. Still the blade is as sharp as the first time it was lifted from the whetstone. At the joints are spaces, and the blade has no thickness. Entering with no thickness where there is space, the blade may move freely where it will: there's plenty of room to move. Thus, after nineteen years, my knife remains

as sharp as it was that first day." This particular translation is found in Hamill and Seaton 1998, chap. 3.

178 Good science too, the modern idea goes: Campbell, O'Rourke, and Slater 2011.

178 Treatment of symptoms, it was argued: Committee on a Framework for Development of a New Taxonomy of Disease 2011, chaps. 1 and 2. This vision of a data-driven, dynamic, taxonomic update in light of molecular and genomic information about patient populations faces all sorts of logistical challenges beyond the conceptual point that I am making here. For reflections on that issue, see S. Green, Carusi, and Hoeyer 2019.

178 using the term "precision medicine": Committee on a Framework for Development of a New Taxonomy of Disease 2011, 12 and 124–25.

179 Health journalists reported: Katsnelson 2013; Timmerman 2013. For a scholarly reflection, see Roden and Tyndale 2013.

179 Now it was precision medicine journals: According to a PubMed search for "precision medicine" or "precision medicines," the year-by-year publications that include these references since 1997 are the following: 1997 (1), 1998 (0), 1999 (0), 2000 (2), 2001 (0), 2002 (1), 2003 (1), 2004 (1), 2005 (1), 2006 (3), 2007 (3), 2008 (24), 2009 (234), 2010 (810), 2011 (1,072), 2012 (1,490), 2013 (1,663).

179 Now universities began adding: Precision medicine programs/centers/institutes can now be found at many universities (particularly those with academic hospitals affiliated).

179 the world that MaryAnne DiCanto found: Attention to "precision medicine" really took off after the National Academies published *Toward Precision Medicine* in 2011, but the term makes its first appearance in its current form in Christensen, Grossman, and Hwang 2009, chaps. 2 and 8. According to Hwang, it was Christensen who pushed for the switch to a term other than "personalized medicine." The coauthors Hwang and Grossman were from medicine and had grown accustomed to using that phrase, while Christensen was from business, and Christensen said truly personalizing medicine would require incorporating individualized information about patients that included more than just molecular-genetic data—things like their mental state, social support, education level, and ability to pay, which "personalized medicine" as it was being used at the time left out. So the group brainstormed and came up with "precision medicine" in its stead. Jason Hwang interview, Oct. 11, 2019. "Stratified medicine" was yet another concept floated around as an alternative to "personalized medicine," but it never took off; see Trusheim, Berndt, and Douglas 2007. The historian Christopher Phillips 2020 reviews some of this history surrounding the rise of "precision medicine" as the term of choice. See also Lemoine 2017.

180 "You think it's going to be precise": Szabo 2018a.

182 only 5–7 percent of the patients benefited: Massard, Michiels, Ferté, et al. 2017; Marquart, Chen, and Prasad 2018.

182 to see what would happen when samples: Kuderer, Burton, Blau, et al. 2017. See also Chae, Davis, Carneiro, et al. 2016. Vinayak Prasad reviews these issues in chapter 8 of Prasad 2020. A related issue pertains to the way that cancer patients

are included in genomic medicine research; they are often lured by the offer to get access to precise medicines tailored to their genomes, when in fact the studies are more often designed to get precise patients that best fit the needs of the study; on this phenomenon, see Dam, Green, Bogicevic, et al. 2022.

182 In 2011, Esmé Savoie: Savoie 2015a and 2019; Marcus 2019.

182 Her mother, Hillary, who documented: Savoie begins describing the results of the *PCDH19* test in her blog, hillarysavoie.com/blog, on March 21, 2013. She documents her trip to the World *PCDH19* Conference in Rome on October 14 and 20, 2013. And she highlights *PCDH19* Epilepsy Awareness Day on November 7, 2013.

183 This second round of testing: Savoie 2015b.

183 At this point, the explanation: Hillary Savoie interview, Nov. 15, 2019.

184 A 2019 study from researchers: SoRelle, Thodeson, Arnold, et al. 2019.

184 "dirty little secret" of DNA: Hillary Savoie interview, Nov. 15, 2019.

184 "It's a testament": UT Southwestern Medical Center 2018.

184 The flaw is in suggesting: Prasad 2016.

185 there is no single target: Anya Plutynski 2018 explores the great complexity surrounding cancer science and cancer medicine. Things get even more complicated when the personalized/precision medicine approach moves outside oncology, to fields like psychiatry. For a reflection on the particular challenges associated with precision psychiatry, see Tabb and Lemoine 2021.

186 If anything, it only exacerbated it: Abola and Prasad 2016; Comfort 2016; Szabo 2017; Caulfield 2018; Marcon, Bieber, and Caulfield 2018; Kaiser 2018; Joyner and Paneth 2019b.

186 Companies offering genetic testing: The language about genomic testing can be found on Foundation Medicine's website for FoundationOne CDx: www.foundationmedicine.com. FoundationOne was one of the genomic tests that MaryAnne DiCanto took.

186 most oncologists continued to say: Cardinal Health 2018. This uncertainty surrounding the likelihood that precision medicine research will generate clinically actionable outcomes is often not conveyed accurately to potential research participants; see Ratcliff, Wong, Jensen, et al. 2021 on how people navigate issues of uncertainty when considering participating in research of this sort.

186 but so too did new journals: The *Journal of Personalized Medicine* was first issued in 2011, and the *Journal of Translational Medicine* added a special section on personalized medicine that same year; *Personalized Medicine Universe* appeared in 2012; in 2017, *Personalized Medicine in Psychiatry,* the *International Journal of Personalized Medicine,* and the *World Journal of Personalized Medicine* all were issued.

186 Academic citations of personalized medicine: According to a PubMed search for "personalized medicine" or "personalized medicines" or "personalised medicine" or "personalised medicines," the year-by-year publication counts from 2009 to 2018 are the following: 2009 (402), 2010 (556), 2011 (673), 2012 (995), 2013 (1,121), 2014 (1,474), 2015 (1,676), 2016 (1,938), 2017 (2,134), 2018 (2,288).

186 University programs offering training: The University of Utah's program in personalized health care, as one example, was launched in 2012.

186 Indeed, the result was about: On personalized medicine being the overarching

concept, under which precision medicine exists, see Pokorska-Bocci, Stewart, Sagoo, et al. 2014, as well as Sudip Parikh's interview of Edward Abrahams at DIA 2019: www.youtube.com. Precision medicine comes first, and then personalized medicine can be found in Christensen, Grossman, and Hwang 2009 as well as Valdes and Yin 2016. Susan Morse 2019 makes the case for precision medicine being about professional-clinical judgments of risk stratification and personalized medicine being about patients equipped with information about their own genes. Muin Khoury 2016 relegates personalized medicine to direct-to-consumer personal genomics, which may or may not be accurate, and equates precision medicine with accurate genomic information that can reliably guide interventions, which may or may not apply to individuals. Marson, Bertuzzo, and Ribeiro 2017 make the case for understanding personalized medicine as treatment direct at symptoms while precision medicine is about the individual in relationship to their environment, lifestyle, and genetics.

187 embracing the two concepts as synonyms: Personalized Medicine Coalition 2017, 5; Konstantinidou, Karaglani, Panagopoulou, et al. 2017, 602; Genetic Support Foundation 2018; Harold 2018; Jackson Laboratory n.d.

188 The president was looking toward: Jorge Contreras tells some of this story at the very end of his enthralling Contreras 2021; President Obama initially told his top science adviser, John Holdren, that he was interested in supporting an initiative at the intersection of genomics, big-data analytics, and public health. Holdren looped in the geneticist Eric Lander and then eventually Collins. Eric Green interview, Aug. 19, 2015.

188 Collins gathered a team: Eric Green interview, Aug. 19, 2015. The core team included Alan Guttmacher, Teri Manolio, Eric Green, Kathy Hudson, Mike Lauer, Stephanie Devaney, and Bill Riley. Collins, meanwhile, wrote up an op-ed for *The Wall Street Journal* making the public case for the value of federal investment in personalized medicine; see Collins 2014a.

189 The room was particularly crowded: Notes from Oct. 3, 2014, Oval Office meeting, as dictated by Collins to Kathy Hudson.

189 "utterly awesome": Collins sent the "utterly awesome" email right after the meeting on October 3, 2014. The "tiger by the tail" comment then came in an email Collins sent the next day, debriefing on the meeting with the president more generally.

190 the group didn't know: Eric Green interview, Aug. 15, 2015; Guttmacher interview, March 18, 2016.

190 "I want the country": "Remarks by the President in State of the Union Address, January 20, 2015," obamawhitehouse.archives.gov.

192 endorsed the idea of recruiting: Information on the convenience sample can be found in the All of Us Research Program Operational Protocol; a pdf of that protocol is available here: allofus.nih.gov.

192 A Genomics Working Group: The Genomics Working Group met in 2017 and produced a report for the All of Us Research Program Advisory Panel, "Considerations Toward a Comprehensive Genomics Strategy"; a pdf of that report can be found here: allofus.nih.gov.

192 Genomic Centers were announced: The Genome Centers, funded with $28.6 million, were announced in 2018: allofus.nih.gov.

192 The program prepared to join forces: Lambert 2019.

192 The FDA, which partnered with the NIH: The FDA's account of precision medicine, equating it with personalized medicine, can be found here: www.fda.gov.

192 Traditional medicine, Collins told reporters: Neergaard 2017.

192 "you as an individual": See the All of Us Program Overview page: allofus.nih.gov.

193 What was once an object of national pride: Hiltzik 2014; Tozzi and Wayne 2014; C. Schmidt 2016.

193 "avoid the ghosts of its past": S. Kaplan 2016.

193 the faults of the National Children's Study: Kaiser 2015.

CHAPTER 8 DISPARITY IN THE GENOME

194 On the afternoon of May 6, 2018: NIH 2018.

196 "Precision Medicine—Prevent Health Disparities": The All of Us Research Program's recruitment website, www.joinallofus.org, had the "Prevent Health Disparities" title for several years, later replacing it with the less promissory "Home | Join All of Us" title.

196 "There are a lot of reasons": Butts had time to only scratch the surface of the horrific history of African Americans' relationships to medical research; see Washington 2006 as well as Skloot 2010 for deeper histories of these biomedical research abuses.

198 the genetics that was played up: This downplaying of the genetics at the center of the All of Us Research Program extended right into the enrollment and consent documents. Of the twenty-six slides that a potential participant had to review in order to sign up and join All of Us in 2018, only one included anything about DNA/genetics.

198 The "All of Us Anthem": You can watch the All of Us Anthem on the NIH Director's Blog here: directorsblog.nih.gov.

198 Two stories from the history: For additional reflections on the limits of the All of Us Research Program to prevent health disparities, see Sankar and Parker 2017, Sabatello and Appelbaum 2017, Rothstein 2017, Cooper and Paneth 2020a, and Parens 2020.

199 The federal researchers were there: Stephanie Stegman tells the story of the NIAMD scientists arriving in Phoenix in Stegman 2010, chap. 3.

200 The desert landscape is harsh: Dimmitt, Comus, and Brewer 2015.

200 as the Gila River cut across: McNamee 2012.

200 the Hohokam people thrived: Woodson 2016.

200 The Akimel O'odham: Wilson 2014; E. C. Wells 2006. The Gila River Indian Community is populated by both the Akimel O'odham (Pima) and the Pee Posh (Maricopa). Colonial terms for Indigenous populations (like "Pima" and "Maricopa") are, for good reason, being phased out in exchange for the communities' preferred

designations. It is not, however, always as simple as just swapping out the colonial term because the old and the new don't necessarily overlap perfectly; "Pima," for example, groups together a number of different Indigenous communities with their own appellations but who all share the O'odham family language. I use "Pima" in this chapter because that is how the scientists who studied the community referenced the population in their studies, and so it is difficult to historically talk about the research and results from that era without employing that term.

201 type 2 diabetes: Arleen Marcia Tuchman 2020 provides a rich history of diabetes and how scientific characterizations of it evolved throughout the twentieth century through various racialized lenses.

201 But the sheer magnitude: Miller, Burch, Bennett, et al. 1965; Schulz and Chaudhari 2015.

201 On June 13, 1966, scientists and administrators: *NIH Record* 1966; Stegman 2010, 76–77.

201 Neel thought diabetes was an evolutionary puzzle: Neel 1962.

203 Something universal had become: Fee 2006; Paradies, Montoya, and Fullerton 2007; Montoya 2011; Gosling, Buckley, Matisoo-Smith, et al. 2015.

203 the NIH scientists in Arizona: Knowler, Pettitt, Bennett, et al. 1983.

203 Neel himself visited: Radin 2017; Tuchman 2020.

203 Malcolm Gladwell's "The Pima Paradox": Gladwell 1998.

204 This decimated their farms: DeJong 1992 and 2009; Booth, Nourian, Weaver, et al. 2017.

204 The federal commodities: Smith-Morris 2006; Joe and Young 1994; California Newsreel with Vital Pictures Inc. 2008, episode 4.

204 A health survey of the Pima: Hrdlička 1908. The environmental source of the health crisis became abundantly clear in the 1990s when the NIH researchers accessed an Indigenous community who were closely related to the Pima but isolated in the remote Sierra Madre Mountains of Mexico; with little exposure to white people and no reliance on federal commodities for nutrition, only three people in the community who were studied could be diagnosed with diabetes. See Ravussin, Valencia, Esparza, et al. 1994 for the original report.

205 The extremely high rate of diabetes: The literature on diabetes among the Pima is immense. Here is just a sample of resources that link up general scientific insights about diabetes to the Pima research: Knowler, Pettitt, Saad, et al. 1990; Pettitt, Nelson, Saad, et al. 1993; Robert G. Nelson, Bennett, Beck, et al. 1996; Baier and Hanson 2004; Franks, Looker, Kobes, et al. 2006; Schulz and Chaudhari 2015; Smith-Morris 2006.

205 The Pima, however, benefited little: Burke, Trinidad, and Schenck 2019; Ozanne and Hales 1998; Cooper and Paneth 2020a; Paradies, Montoya, and Fullerton 2007.

205 Neel admitted that his thrifty genotype: Neel, Weder, and Julius 1998; Neel 1999.

205 Still, the damage was done: Paradies, Montoya, and Fullerton 2007; Burke, Trinidad, and Schenck 2019; McDermott 1998; Ferreira and Lang 2006. Puneet Chawla Sahota 2012 provides an ethnographic insight into how Indigenous peoples them-

selves incorporated the thrifty genotype concept into their own narratives about being at risk for diabetes.

206 The community collectively decided: Smith-Morris 2006; Cooper and Paneth 2020a.

206 What is well understood: There are a tremendous number of resources available on the social/environmental nature of health disparities alongside the unanswered questions surrounding them. As an entry into that literature, consider the following: D. S. Jones 2005; Sarche and Spicer 2008; Braveman, Egerter, and Williams 2011; Chowkwanyun 2011; Hoberman 2012; Hicken, Gee, Morenoff, et al. 2012; Braveman and Gottlieb 2014; Matthew 2015; Walker, Williams, and Egede 2016; Barr 2019; Hammonds and Reverby 2019; Causadius and Korous 2019; Payne-Sturges, Gee, and Cory-Slechta 2021.

207 There is little reason to believe: Dorothy Roberts and Wylie Burke have both written extensively about the dangers associated with seeking to alleviate the harm of racial health disparities by focusing on genetic differences. See D. Roberts 2008, 2011a, 2011b, 2012; Burke, Trinidad, and Schenck 2019; McGlone, Blacksher, and Burke 2017.

207 For sickle-cell disease: The historian to read on all things sickle-cell disease is Keith Wailoo, who has authored or co-authored several fascinating books on the history of scientific studies of the disease, as well as the comparative history between sickle cell and cystic fibrosis. See, for example, Wailoo 2001, Wailoo and Pemberton 2006, and Wailoo 2017.

207 Cystic fibrosis is equally debilitating: Doershuk 2001.

208 the cystic fibrosis transmembrane conductance regulator gene: Remember from chapter 1 that this was one of the genes Francis Collins helped discover in the early 1990s, launching his meteoric rise to fame.

208 sickle cell became labeled: Pauling, Itano, Singer, et al. 1949. Wailoo and Pemberton 2006 tell the comparative histories of cystic fibrosis and sickle cell in chapters 2 and 3 of their book. On the specific discovery by Pauling and his team (and its place in the broader history of research on sickle cell), see also Gormley 2007 and S. D. Feldman and Tauber 1997.

209 John Strouse, a hematologist: Strouse, Lobner, Lanzkron, et al. 2013; Farooq and Strouse 2018; Farooq, Mogayzel, Lanzkron, et al. 2020. Robert Scott, a Virginia Commonwealth University hematologist, actually raised alarms about the underfunding of sickle-cell disease back in 1970 in Scott 1970a and 1970b. See also L. A. Smith, Oyeku, Homer, et al. 2006; Grosse, Schechter, Kulkarni, et al. 2009.

209 Those differences in monetary support: Butler 2015; Gold 2017; Lanzkron 2018; Carroll 2019.

210 Nearly half a million people: Denny 2020. The problems identified here are exacerbated by the fact that All of Us utilizes what is called "broad consent" when enrolling participants. Traditionally, researchers who were interested in enrolling potential research participants in a study had to tell the people what the study was about (e.g., the problem it aimed to address, the potential impact, etc.); that way, the people could decide whether they wanted to participate and add their

own data to the scientific effort. With broad consent, however, potential research participants are asked only to add their data to a repository, and then they have no control over which studies accessing that data do or do not utilize their personal data. So people can be lured to participate in the scientific effort with the promise of one kind of research and, once participants' data is included in the repository, the research can go in a completely different direction without the participants having any idea about that new trajectory.

210 the same frustrating situation as the Pima: Members of the Native BioData Consortium have repeatedly warned about the risks posed to Indigenous communities by these large-scale genomic data grabs, as well as the reasons to be skeptical about the extent to which the research using the data will address the health crises affecting those communities; see, for example, Tsosie, Yracheta, and Dickenson 2019; Fox 2020; Tsosie, Begay, Fox, et al. 2020; Hudson, Garrison, Sterling, et al. 2020; Tsosie, Yracheta, Kolopenuk, et al. 2021; and Tsosie, Fox, and Yracheta 2021. The Consortium also provides a wealth of resources on issues regarding Indigenous data sovereignty, data stewardship, and tribal-centered approaches to consent at nativebio.org.

210 Those insights, however, will be of little help: The medical ethicist Sandra Soo-Jin Lee presciently predicted many of these concerns when personalized medicine was first taking shape in the field of pharmacogenomics; see Lee 2009.

CHAPTER 9 THE "GLEEVEC SCENARIO"

211 Chronic myelogenous leukemia: The National Cancer Institute provides ongoing updates on chronic myelogenous leukemia statistics at seer.cancer.gov.

211 brand name Gleevec: Imatinib was branded as "Glivec" everywhere except in the United States, where it was instead named "Gleevec" (in order to differentiate it from a different, similarly spelled drug already on the market).

212 Indeed, oncologists found that Gleevec: Brian Druker describes the pre- versus post-Gleevec treatment landscape in his interview with Dreifus 2009.

212 It even played a role: *Washington Post* 2002.

212 "Before Gleevec" and "After Gleevec": Monmaney 2011.

212 health care transitioned into the age: Wapner 2013.

212 He hoped the conversation: Druker interview, April 30, 2021.

213 So he phoned the FDA: Ibid.; Wapner 2013, 163; Monmaney 2011.

214 The discovery of that chromosomal abnormality: Nowell 2007; Wapner 2013. Chromosome abnormalities are standardly named after the city where the discovery is made.

214 The second piece of the puzzle: Waalen 2001.

214 Cancer treatment at the time: Druker used the baseball bat metaphor in his interview with Dreifus 2009.

214 STI-571 was the best contender: Dreifus 2009; Monmaney 2011; Wapner 2013, 168.

215 Every participant responded positively: The tremendous response of all the

patients gets referenced in just about every story surrounding Gleevec; it's the first indication that Druker was onto something very big.

215 "I'd never written it up": Dreifus 2009.

215 Suzan McNamara: Monmaney 2011; Wapner 2013, chap. 28.

215 The results were genuinely astounding: Druker, Talpaz, Resta, et al. 2001. The article concludes with a nod to the great promise of Gleevec and the future of personalized medicine: "These results show that the BCR-ABL tyrosine kinase is critical to the development of CML and demonstrate the potential for the development of anticancer drugs based on the specific molecular abnormality in a human cancer."

216 Ten weeks later the agency approved imatinib: Okie 2001.

216 Vasella joined Tommy Thompson: Wade 2001.

216 Imatinib was described variously: Billingsley 2001; Peck 2001; Atkins and Gershell 2002.

216 The tiny pills appeared on the cover: *Time,* May 28, 2001.

216 Vasella admitted the price was steep: Vasella and Slater 2003.

216 Novartis typically invested: Wade 2001.

216 the price could come down: Carolyn Y. Johnson provides an in-depth analysis of Gleevec's rising price over the years for *The Washington Post* in C. Y. Johnson 2016; she flags and directs readers to the 2001 company newsletter where the CEO, Vasella, indicated that the cost of imatinib could go down if the patient population expands.

216 Around 2006, the price began climbing: Ibid.

217 a petition to Change.org: The Change.org petition (and supporters' comments on it) can be found here: www.change.org.

217 "unsustainable prices of cancer drugs": Experts in Chronic Myelogenous Leukemia 2013.

217 "Brian, if we don't do this, who will?": Druker interview, April 30, 2021.

217 the company filed for a series of extensions: Conti 2013.

218 Novartis utilized a pay-for-delay strategy: Falconi 2014.

218 Novartis marketing gradually directed: C. Y. Johnson 2016; Freudenberg 2021, chap. 4.

218 list price of about $5,000 a month: Cole and Dusetzina 2018; Tessema, Kesselheim, and Sinha 2020; Katrina A. Fischer, Robert E. Miller, and John A. Glaspy describe this phenomenon as "price stickiness" in their 2020 piece for *JAMA Oncology.*

218 The science advertised as personalized medicine: The most accessible avenue into the pricing of specialty drugs like Gleevec even after a generic arrives is to be found in Joshua Cohen's 2018 article for *Forbes.*

218 finally dipped to a reasonable rate: Roxanne Nelson 2019.

218 roughly $50 billion in global sales: A congressional investigation announced Novartis's financial gains from Gleevec between 2009 and 2019: Committee on Oversight and Reform, U.S. House of Representatives 2020.

218 an extremely pricey diagnosis: Wilkes, Lyman, Doody, et al. 2019.

218 steady arrival of shockingly expensive drugs: Henk, Winestone, Wilkes, et al. 2020; Lyman and Henk 2020; Mapes 2020.

218 Erin Havel: Havel recounts her experience with chronic myelogenous leukemia and the exhausting maze trying to get reliable access to Gleevec in Havel 2012. Her story has also been highlighted by a number of health reporters discussing the financial toxicity associated with chronic myelogenous leukemia care; see, for example, Mapes 2015 and Radcliffe 2019.

219 "We're not going to do your treatment today": Havel 2012, chap. 7.

219 she was comforted, however: Havel interview, April 22, 2021.

219 Gleevec, to Havel's shock: Havel 2012, chap. 8.

219 The diagnosis of leukemia: Ibid., chap. 9.

220 To her disappointment and embarrassment: Ibid., chap. 19.

220 attendees shared stories of "drug smuggling": Ibid., 199.

220 many people suffering from the disease: Dusetzina, Winn, Abel, et al. 2014; Winn, Keating, and Dusetzina 2016.

221 the costs of oncological therapies: Carrera and Olver 2015; Huntington, Davidoff, and Gross 2019.

221 Those tests can cost hundreds: Schilsky 2020.

221 A recent survey of oncologists: Cardinal Health 2018.

221 financial toxicity: Zafar and Abernathy 2013; Morrison 2015; Carrera, Kantarjian, and Blinder 2018; Gilligan, Alberts, Roe, et al. 2018; Ramsey, Blough, Kirchhoff, et al. 2013; Ramsey, Bansal, Fedorenko, et al. 2016.

222 they spent about half a million dollars: Primiano interview, Nov. 11, 2019.

222 Kilmer shared her heartbreaking plight: Szabo 2018b.

222 Havel, fortunately, eventually landed on her feet: Havel interview, April 22, 2021.

223 Havel warned in an essay: Havel 2015.

223 "And if you think that sounds like science fiction": Collins begins discussing Gleevec at around the three-hour fourteen-minute mark of the NIH videocast: videocast.nih.gov.

225 Studies of Parkinson's disease: von Linstow, Gan-Or, and Brundin 2020.

225 Asthma genetics followed a similar trajectory: Meyers, Bleecker, Holloway, et al. 2014.

225 The genetics of diabetes: Fuchsberger, Flannick, Teslovich, et al. 2016; McCarthy 2017.

225 when that assumption proved erroneous: Another way to put this point is to say that the discoveries of the "major contributing genes" were supposed to pinpoint the molecular mechanisms that gave rise to the common diseases under investigation, which could then be intervened upon to prevent the development of them. Instead, the genotyping and sequencing efforts turned up common variants with little insights about the molecular mechanisms or turned up rare variants that did pinpoint molecular mechanisms but which only applied to a very small number of people. There is a very large philosophical literature on how scientists discover and explain the natural world by way of elucidating mechanisms; for a sample of the most influential works, see Darden 2006, Bechtel 2006, Craver 2007, and Glennan 2017.

225 Almost all people with Parkinson's: Gasser 2015.

225 Asthma prediction based on: Hernandez-Pacheco, Pino-Yanes, and Flores 2019.

226 For type 2 diabetes: Vassey, Hivert, and Porneala, et al. 2014.

226 Illnesses like asthma and type 2 diabetes: El-Husseini, Gosens, Dekker, et al. 2020; Galeone, Scelfo, Bertolini, et al. 2018; Chung, Erion, Florez, et al. 2020.

226 Pharmaceutical companies like Pfizer: Schneider and Alcalay 2020.

227 Parkinson's disease is among the fastest-growing: Savica, Grossardt, Bower, et al. 2016; Dorsey, Sherer, Okun, et al. 2018; Marras, Beck, Bower, et al. 2018; Dorsey, Elbaz, Nichols, et al. 2018; Yang, Hamilton, Kopil, et al. 2020.

227 Asthma rates also spiked: Hegner 2000; Caggiano 2019.

227 Diabetes in America followed: Chow, Foster, Gonzalez, et al. 2012; Pettitt, Talton, Dabalea, et al. 2014; Menke, Casagrande, Geiss, et al. 2015; Divers, Mayer-Davis, Lawrence, et al. 2020.

228 Industrial solvents, lead, air pollution: Goldman, Quinlan, Ross, et al. 2011; Goldman 2014; Gatto, Cockburn, Bronstein, et al. 2009.

228 Increasing rates of asthma: Greenwood 2011; Guarnieri and Balmes 2014; Dharmage, Perret, and Custovic 2019; Keet, McCormack, Pollack, et al. 2015.

228 Exposure to mothers' gestational diabetes: Dabalea, Mayer-Davis, Lamicchane, et al. 2008; Page, Luo, Wang, et al. 2019. The fetal environment is known to have a tremendous impact on the developmental trajectory of a child; the extent to which exposures during that period can increase risk of chronic illness later in life—the fetal origin hypothesis—is an ongoing focus of research. See Almond and Currie 2011 for an introduction to this literature.

228 The built environments: Kolb and Martin 2017; Dendup, Feng, Clingan, et al. 2018.

Bibliography

Abola, Matthew V., and Vinay Prasad. 2016. "The Use of Superlatives in Cancer Research." *JAMA Oncology* 2:139–41.

Abraham, John. 1995. *Science, Politics, and the Pharmaceutical Industry: Controversy and Bias in Drug Regulation.* London: University College London Press.

Abrahams, Edward. 2005. "The Personalized Medicine Coalition—8 Months On." *Personalized Medicine* 2:193–95.

———. 2008. "Letter from the Executive Director." *Personalized Medicine* 6:131–32.

Abrahams, Edward, Geoffrey S. Ginsburg, and Mike Silver. 2005. "The Personalized Medicine Coalition: Goals and Strategies." *American Journal of Pharmacogenomics* 5:345–55.

Abramson, Harold. 1944. "Accidental Mechanical Suffocation in Infants." *Journal of Pediatrics* 25:404–13.

Akdis, Cezmi A., and Zuhair K. Ballas. 2016. "Precision Medicine and Precision Health: Building Blocks to Foster a Revolutionary Health Care Model." *Journal of Allergy and Clinical Immunology* 137:1359–61.

Albert, Keith J., and David R. Walt. 2000. "High-Speed Fluorescence Detection of Explosives-Like Vapors." *Analytical Chemistry* 72:1947–55.

Albert, Keith J., M. L. Myrick, Steve B. Brown, et al. 2001. "Field-Deployable Sniffer for 2,4-Dinitrotoluene Detection." *Environmental Science & Technology* 35: 3193–3200.

Albert, Keith J., David R. Walt, Daljeet S. Gill, et al. 2001. "Optical Multibead Arrays for Simple and Complex Odor Discrimination." *Analytical Chemistry* 73:2501–8.

Allison, Malorye. 2008. "Industry Welcomes Genetic Information Nondiscrimination Act." *Nature Biotechnology* 26:596–97.

Almond, Douglas, and Janet Currie. 2011. "Killing Me Softly: The Fetal Origins Hypothesis." *Journal of Economic Perspectives* 25:153–72.

Alving, Alf S., Branch Craige Jr., Theodore N. Pullman, et al. 1948. "Procedures Used at Stateville Penitentiary for the Testing of Potential Antimalarial Agents." *Journal of Clinical Investigation* 27:2–5.

Anand, Geeta. 2001. "Big Drug Companies Try to Delay 'Personalized' Drug Regimens." *Wall Street Journal,* June 18, 2001. www.wsj.com.

Andrews, Richard L. 2006. *Managing the Environment, Managing Ourselves: A His-*

tory of American Environmental Policy. 2nd ed. New Haven, Conn.: Yale University Press.

———. 2010. "The EPA at 40: An Historical Perspective." *Duke Environmental Law and Policy Forum* 21:223–58.

Anthes, Emily, 2008. "E-noses Could Make Diseases Something to Sniff At." *Scientific American,* Jan. 11, 2008. www.scientificamerican.com.

Armstrong, Katrina. 2017. "Equity in Precision Medicine: Is It Within Our Reach?" *Journal of the National Comprehensive Cancer Network* 15:422–23.

Arnold, Bettina. 1990. "The Past as Propaganda: Totalitarian Archaeology in Nazi Germany." *Antiquity* 64:464–78.

———. 2006. "'*Arierdämmerung*': Race and Archaeology in Nazi Germany." *World Archaeology* 38:8–31.

Asbury, Kathryn, and Robert Plomin. 2013. *G Is for Genes: The Impact of Genetics on Education and Achievement.* Malden, Mass.: Wiley Blackwell.

Atkins, Joshua H., and Leland J. Gershell. 2002. "Selective Anticancer Drugs." *Nature Reviews Drug Discovery* 1:491–92.

Bach, Peter B., Laura D. Cramer, Joan L. Warren, et al. 1999. "Racial Differences in the Treatment of Early-Stage Lung Cancer." *New England Journal of Medicine* 341:1198–205.

Baier, Leslie J., and Robert L. Hanson. 2004. "Genetic Studies of the Etiology of Type 2 Diabetes in Pima Indians: Hunting for Pieces of a Complicated Puzzle." *Diabetes* 53:1181–86.

Bailey, David G., J. Malcolm, O. Arnold, et al. 1998. "Grapefruit Juice–Drug Interactions." *British Journal of Clinical Pharmacology* 46:101–10.

Barnard, Steven M., and David R. Walt. 1991. "Fiber-Optic Organic Vapor Sensor." *Environmental Science and Technology* 25:1301–4.

Barr, Donald A. 2019. *Health Disparities in the United States: Social Class, Race, Ethnicity, and the Social Determinants of Health.* 3rd ed. Baltimore: Johns Hopkins University Press.

Bates, Gillian P. 2005. "The Molecular Genetics of Huntington Disease—a History." *Nature Reviews Genetics* 6:766–73.

Baudou, Evert. 2005. "Kossinna Meets the Nordic Archaeologists." *Current Swedish Archaeology* 13:121–39.

Bayer, Ronald, and Sandro Galea. 2015. "Public Health in the Precision-Medicine Era." *New England Journal of Medicine* 373:499–501.

BBC. 2014. "Pfizer: The Making of a Global Drugs Giant." May 14, 2014. www.bbc.com.

Bechtel, William. 2006. *Discovering Cell Mechanisms: The Creation of Modern Cell Biology.* Cambridge, U.K.: Cambridge University Press.

Berman, Elizabeth Popp. 2012. *Creating the Market University: How Academic Science Became an Economic Engine.* Princeton: Princeton University Press.

Berrett, Dan. 2007. "Got Radon?" *Pocono Record,* Jan. 21, 2007. www.poconorecord.com.

Billingsley, Janice. 2001. "A 'Miracle' Drug for Leukemia." HealthDay, May 10, 2001. consumer.healthday.com.

Blackburn, Maria. 2005. "Lessons from the Womb." *Johns Hopkins Magazine* 57. pages .jh.edu.

Blankstein, Sarah. 2014. "Pharmacogenomics: History, Barriers, and Regulatory Solutions." *Food and Drug Law Journal* 69:273–314.

Bloom, Floyd E., and Mark A. Randolph, eds. 1990. *Funding Health Sciences Research: A Strategy to Restore Balance.* Washington, D.C.: National Academies Press.

Bochnak, Bob. 2002. "Doctoral Student Works to Rid the World of Land Mines." *Tufts Journal,* Feb. 2002. tuftsjournal.tufts.edu.

Booth, Clayton, Maziar M. Nourian, Shannon Weaver, Bethany Gull, and Akiko Kamimura. 2017. "Policy and Social Factors Influencing Diabetes Among Pima Indians in Arizona, USA." *Public Policy and Administration Research* 7:35–39.

Bora, Chandramita. 2018. "The Fitbit Story: How It Scripted Wearable Tech's Biggest Success Story." Tech Story, Dec. 18, 2018. techstory.in.

Bottles, Kent. 2001. "A Revolution in Genetics: Changing Medicine, Changing Lives." *Physician Executive* 27:58–63.

Bouregy, Susan, Elena L. Grigorenko, Stephen R. Latham, et al., eds. 2017. *Genetics, Ethics, and Education.* Cambridge, U.K.: Cambridge University Press.

Bradley, Cathy J., K. Robin Yabroff, Angela B. Mariotta, et al. 2017. "Antineoplastic Treatment of Advanced-Stage Non-Small-Cell Lung Cancer: Treatment, Survival, and Spending (2000 to 2011)." *Journal of Clinical Oncology* 35:529–36.

Braveman, Paula, Susan Egerter, and David R. Williams. 2011. "The Social Determinants of Health: Coming of Age." *Annual Review of Public Health* 32:381–98.

Braveman, Paula, and Laura Gottlieb. 2014. "The Social Determinants of Health: It's Time to Consider the Causes of the Causes." *Public Health Reports* 129:19–31.

Brawley, Otis W., Thomas J. Glynn, Fadlo R. Khuri, et al. 2013. "The First Surgeon General's Report on Smoking and Health: The 50th Anniversary." *CA: A Cancer Journal for Clinicians* 64:5–8.

Bringing the Genome to You. 2003. National Museum of Natural History. Symposium, April 15, 2003. www.genome.gov.

Bullard, Robert D. 1990. *Dumping in Dixie: Race, Class, and Environmental Quality.* Boulder, Colo.: Westview Press.

———. 1993. *Confronting Environmental Racism: Voices from the Grassroots.* Boston: South End Press.

Burke, Wylie, Susan Brown Trinidad, and David Schenck. 2019. "Can Precision Medicine Reduce the Burden of Diabetes?" *Ethnicity and Disease* 29:669–74.

Burns, Alexander. 2008. "Obama Announces Transition-Team Staff." *Politico,* Nov. 6, 2008. www.politico.com.

Business Wire. 2004. "Perlegen Sciences Teams with Pfizer on Metabolic Syndrome Research." Jan. 7, 2004. www.businesswire.com.

Butler, Kiera. 2015. "One Disease Hits Mostly People of Color. One Mostly Whites. Which One Gets Billions in Funding?" *Mother Jones,* May 4, 2015. www.motherjones.com.

Caggiano, Regina. 2019. "The Asthma Epidemic Isn't Over." *Brown Political Review,* Nov. 13, 2019. brownpoliticalreview.org.

California Newsreel with Vital Pictures Inc. 2008. *Unnatural Causes: Is Inequality Making Us Sick?* Episode 4: "Bad Sugar." unnaturalcauses.org.

Callaway, Ewen. 2018. "The Battle for Common Ground." *Nature* 555:573–76.

Campbell, Joseph Keim, Michael O'Rourke, and Matthew H. Slater, eds. 2011. *Carving*

Nature at Its Joints: Natural Kinds in Metaphysics and Science. Cambridge, Mass.: MIT Press.

Cann, Rebecca L., Mark Stoneking, and Allan C. Wilson. 1987. "Mitochondrial DNA and Human Evolution." *Nature* 325:31–36.

Cardinal Health. 2018. "Views on Precision Medicine and Genomic Testing Practices from Specialty Physicians Nationwide." *Oncology Insights,* June 2018, 1–12.

Carrera, Pricivel M., Hagop M. Kantarjian, and Victoria S. Blinder. 2018. "The Financial Burden of Distress of Patients with Cancer: Understanding and Stepping-Up Action on the Financial Toxicity of Cancer Treatment." *CA: A Cancer Journal for Clinicians* 68:153–65.

Carrera, Pricivel M., and Ian Olver. 2015. "The Financial Hazard of Personalized Medicine and Supportive Care." *Supportive Care in Cancer* 23:3399–401.

Carroll, Aaron E. 2019. "Sickle Cell Disease Still Tends to Be Overlooked." *New York Times,* Aug. 5, 2019. www.nytimes.com.

Carson, Paul E., C. Larkin Flanagan, C. E. Ickes, et al. 1956. "Enzymatic Deficiency in Primaquine-Sensitive Erythrocytes." *Science* 124:484–85.

Carson, Rachel. 1962. *Silent Spring.* New York: Houghton Mifflin.

Caulfield, Timothy. 2018. "Spinning the Genome: Why Science Hype Matters." *Perspectives in Biology and Medicine* 61:560–71.

Causadius, José M., and Kevin M. Korous. 2019. "Racial Discrimination in the United States: A National Health Crisis That Demands a National Health Solution." *Journal of Adolescent Health* 64:147–48.

Causey, Mike. 1999. "Government Intern Program, by Another Name, Might Incite Less Mirth." *Washington Post,* March 9, 1999.

Chae, Young Kwang, Andrew A. Davis, Benedito A. Carneiro, et al. 2016. "Concordance Between Genomic Alterations Assessed by Next-Generation Sequencing in Tumor Tissue or Circulating Cell-Free DNA." *Oncotarget* 7:65364–73.

Chan, Mi-Anne. 2016. "11 True Stories That Will Make You Think Twice About Your Lipstick." Yahoo! Lifestyle, Oct. 29, 2016. www.yahoo.com.

Chase, Lauren. 2020. "The 10 Most Expensive Drugs in the U.S., Period." GoodRx, Aug. 25, 2020. www.goodrx.com.

"Child Health." 2002. *Washington Journal,* C-SPAN, Nov. 27, 2002. www.c-span.org.

Children's Health Act of 2000. 2000. 106th Cong., H.R. 4365. www.govinfo.gov.

Choudhury, Narmeen. 2015. "Woman with Metastatic Breast Cancer Talks Struggles and Triumphs." PIX11, Oct. 17, 2015. pix11.com.

Chow, Edward A., Henry Foster, Victor Gonzalez, et al. 2012. "The Disparate Impact of Diabetes on Racial/Ethnic Minority Populations." *Clinical Diabetes* 30:130–33.

Chowkwanyun, Merlin. 2011. "The Strange Disappearance of History from Racial Health Disparities Research." *Du Bois Review: Social Science Research on Race* 8:253–70.

Chowkwanyun, Merlin, Ronald Bayer, and Sandro Galea. 2018. "'Precision' Public Health—Between Novelty and Hype." *New England Journal of Medicine* 379:1398–400.

Christensen, Clayton M., Jerome H. Grossman, and Jason Hwang. 2009. *The Innovator's Prescription: A Disruptive Solution for Health Care.* New York: McGraw-Hill.

Chung, Wendy K., Karel Erion, Jose C. Florez, et al. 2020. "Precision Medicine in Diabetes: A Consensus Report from the American Diabetes Association (ADA) and the European Association for the Study of Diabetes (EASD)." *Diabetes Care* 43:1617–35.

Clinton, Bill. 2000. "Statement on Signing the Children's Health Act of 2000." Oct. 17, 2000. www.gpo.gov.

Cohen, Joshua. 2018. "The Curious Case of Gleevec Pricing." *Forbes*, Sept. 12, 2018. www.forbes.com.

Cohen, Martin H., John R. Johnson, Yeh-Fong Chen, et al. 2005. "FDA Drug Approval Summary: Erlotinib (Tarceva®) Tablets." *Oncologist* 10:461–66.

Cole, Ashley, and Stacie B. Dusetzina. 2018. "Generic Price Competition for Specialty Drugs: Too Little, Too Late?" *Health Affairs* 37:738–42.

Collins, Francis S. 2006. *The Language of God: A Scientist Presents Evidence for Faith.* New York: Free Press.

———. 2010a. "Opportunities for Research and NIH." *Science* 327:36–37.

———. 2010b. "Has the Revolution Arrived?" *Nature* 464:674–75.

———. 2010c. *The Language of Life: DNA and the Revolution in Personalized Medicine.* New York: HarperCollins.

———. 2010d. "A Genome Story: 10th Anniversary." *Scientific American*, June 25, 2010. blogs.scientificamerican.com.

———. 2014a. "Francis Collins Says Medicine in the Future Will Be Tailored to Your Genes." *Wall Street Journal*, July 7, 2014. wsj.com.

———. 2014b. "Statement on the National Children's Study." www.nih.gov.

Collins, Francis S., and Harold Varmus. 2015. "A New Initiative on Precision Medicine." *New England Journal of Medicine* 372:793–95.

Colten, Craig E., and Peter N. Skinner. 1996. *The Road to Love Canal: Managing Industrial Waste Before EPA.* Austin: University of Texas Press.

Comfort, Nathaniel. 2009. "The Prisoner as Model Organism: Malaria Research at Stateville Penitentiary." *Studies in the History and Philosophy of Biological and Biomedical Sciences* 40:190–203.

———. 2012. *The Science of Human Perfection: How Genes Became the Heart of American Medicine.* New Haven, Conn.: Yale University Press.

———. 2016. "The Overhyping of Precision Medicine." *Atlantic*, Dec. 12, 2016. www.theatlantic.com.

Committee on a Framework for Development of a New Taxonomy of Disease. 2011. *Toward Precision Medicine: Building a Knowledge Network for Biomedical Research and a New Taxonomy of Disease.* Washington, D.C.: National Academies Press.

Committee on Oversight and Reform, U.S. House of Representatives. 2020. "Drug Pricing Investigation: Novartis—Gleevec." Staff Report, Committee on Oversight and Reform U.S. House of Representatives, Oct. 2020. oversight.house.gov.

Committee on Pesticides in the Diets of Infants and Children. 1993. *Pesticides in the Diets of Infants and Children.* Washington, D.C.: National Academies Press. www.nap.edu.

Commonwealth Fund. 2006. "Harkin Bemoans Plan to Shelve National Children's Study." May 30, 2006. www.commonwealthfund.org.

Connelly, Emily, and Stacie Propst. 2006. "2005 Investment in U.S. Health Research." Research!America. www.researchamerica.org.

———. 2007. "2006 Investment in U.S. Health Research." Research!America. www.researchamerica.org.

Connor, Steve. 1999. "Genetic Blueprint to Predict Illness." *Independent,* April 15, 1999. www.independent.co.uk.

Conti, Rena M. 2013. "Why Are Cancer Drugs Commonly the Target of Schemes to Extend Patent Exclusivity?" *Health Affairs,* Dec. 4, 2013. www.healthaffairs.org.

Contreras, Jorge. 2021. *The Genome Defense: Inside the Epic Legal Battle to Determine Who Owns Your DNA.* Chapel Hill, N.C.: Algonquin Books.

Cook-Deegan, Robert. 1995. *The Gene Wars: Science, Politics, and the Human Genome.* New York: W. W. Norton.

Cooper, Richard, and Nigel Paneth. 2020a. "Will Precision Medicine Lead to a Healthier Population?" *Issues in Science and Technology* 36 (Winter). issues.org.

———. 2020b. "Precision Medicine: Course Correction Urgently Needed." *STAT,* March 3, 2020. www.statnews.com.

Corcos, Alain F. 1984. "Reproduction and Hereditary Beliefs of the Hindus Based on Their Sacred Books." *Journal of Heredity* 75:152–54.

Cortez, Amanda Daniela, Deborah A. Bolnick, George Nichols, et al. 2021. "An Ethical Crisis in Ancient DNA Research: Insights from the Chaco Canyon Controversy as a Case Study." *Journal of Social Archaeology,* March 3, 2021.

Corvin, Aiden, Nick Craddock, and Patrick F. Sullivan. 2010. "Genome-Wide Association Studies: A Primer." *Psychological Medicine* 40:1063–77.

Cowan, Ruth Schwartz. 2008. *Heredity and Hope: The Case for Genetic Screening.* Cambridge, Mass.: Harvard University Press.

Cowgill, Brittney. 2018. *Rest Uneasy: Sudden Infant Death Syndrome in Twentieth-Century America.* New Brunswick, N.J.: Rutgers University Press.

Craver, Carl F. 2007. *Explaining the Brain: Mechanisms and the Mosaic Unity of Neuroscience.* Oxford: Oxford University Press.

Czarnik, Anthony. 2019. "Illumina—the Origin Story." Aug. 20, 2019. www.linkedin.com.

Dabalea, Dana, Elizabeth J. Mayer-Davis, Archana P. Lamicchane, et al. 2008. "Association of Intrauterine Exposure to Maternal Diabetes and Obesity with Type 2 Diabetes in Youth." *Diabetes Care* 31:1422–26.

Dalton, Rex. 2010. "Ancient DNA Set to Rewrite Human History." *Nature* 465:148–49.

Daly, A. K. 2007. "Individualized Drug Therapy." *Current Opinion in Drug Discovery and Development* 10:29–36.

Dam, Mie S., Sara Green, Ivana Bogicevic, et al. 2022. "Precision Patients: Selection Practices and Moral Pathfinding in Experimental Oncology." *Sociology of Health and Illness* 44:345–59. doi.org/10.1111/1467-9566.13424.

D'Ambrosia, Robert D. 2018. "Precision Medicine: A New Frontier in Spine Surgery." *Orthopedics* 39:75–76.

Danielson, P. B. 2002. "The Cytochrome P450 Superfamily: Biochemistry, Evolution, and Drug Metabolism in Humans." *Current Drug Metabolism* 3:561–97.

Dankwa-Mullan, Irene, Jonca Bull, and Francisco Sy. 2015. "Precision Medicine and Health Disparities: Advancing the Science of Individualizing Patient Care." *American Journal of Public Health* (Supplement 3): S368.

Darden, Lindley. 1991. *Theory Change in Science: Strategies from Mendelian Genetics.* New York: Oxford University Press.

———. 2006. *Reasoning in Biological Discoveries: Mechanism, Interfield Relations, and Anomaly Resolution.* New York: Cambridge University Press.

Das, Reenita. 2017. "Drug Industry Bets Big on Precision Medicine: Five Trends Shaping Care Delivery." *Forbes,* March 8, 2017. www.forbes.com.

Davies, David P. 1984. "Cot Death in Hong Kong: A Rare Problem?" *Lancet* 326:1346–49.

Davies, Kevin. 2001. *Cracking the Genome: Inside the Race to Unlock Human DNA.* New York: Free Press.

Davis, Oliver S. P., Lee M. Butcher, Sophia J. Docherty, et al. 2010. "A Three-Stage Genome-Wide Association Study of General Cognitive Ability: Hunting the Small Effects." *Behavior Genetics* 40:759–67.

de Groot, Patricia M., Carol C. Wu, Brett W. Carter, et al. 2018. "The Epidemiology of Lung Cancer." *Translational Lung Cancer Research* 7:220–33.

DeJong, David H. 1992. " 'See the New Country': The Removal Controversy and Pima-Maricopa Water Rights, 1869–1879." *Journal of Arizona History* 33:367–96.

———. 2009. *Stealing the Gila: The Pima Agricultural Economy and Water Deprivation, 1848–1921.* Tucson: University of Arizona Press.

Dendup, Tashi, Xiaoqi Feng, Stephanie Clingan, et al. 2018. "Environmental Risk Factors for Developing Type 2 Diabetes Mellitus: A Systematic Review." *International Journal of Environmental Research and Public Health* 15:78.

Denny, Josh. 2020. "All of Us Research Program Begins Beta Testing of Data Platform." May 27, 2020. allofus.nih.gov.

Deseret News. 2005. "Disease 'Map' Is Unveiled in Utah." Oct. 27, 2005. www.deseretnews.com.

Devine, Robert S. 2004. *Bush Versus the Environment.* New York: Anchor Books.

Dharmage, Shyamali, Jennifer L. Perret, and Adnan Custovic. 2019. "Epidemiology of Asthma in Children and Adults." *Frontiers in Pediatrics* 7:246.

Díaz-Barriga, Fernando, Lilia Batres, Jaqueline Calderón, et al. 1997. "The El Paso Smelter 20 Years Later: Residual Impact on Mexican Children." *Environmental Research* 74:11–16.

Dickenson, Donna. 2013. *Me Medicine vs. We Medicine: Reclaiming Biotechnology for the Common Good.* New York: Columbia University Press.

Dickinson, Todd A., Karri L. Michael, John S. Kauer, et al. 1999. "Convergent, Self-Encoded Bead Sensor Array in the Design of an Artificial Nose." *Analytical Chemistry* 71:2192–98.

Dickinson, Todd A., Joel White, John S. Kauer, et al. 1996. "A Chemical-Detecting System Based on a Cross-Reactive Optical Sensor Array." *Nature* 382:697–700.

Dimmitt, Mark A., Patricia Wentworth Comus, and Linda M. Brewer, eds. 2015. *A Natural History of the Sonoran Desert.* Tucson: Arizona-Sonora Desert Museum Press.

Divaris, Kimon. 2017. "Fundamentals of Precision Medicine." *Compendium of Continuing Education in Dentistry* 38:30–32.

Divers, Jasmin, Elizabeth J. Mayer-Davis, Jean M. Lawrence, et al. 2020. "Trends in Incidence of Type 1 and Type 2 Diabetes Among Youth—Selected Counties and Indian

Reservations, United States, 2002–2015." *Morbidity and Mortality Weekly Report* 69:161–65.

Doershuk, Carl F., ed. 2001. *Cystic Fibrosis in the 20th Century: People, Events, and Progress.* Cleveland: AM.

Dolan, Brian. 2009. "Illumina Demos Concept iPhone App for Genetic Data Sharing." Mobi Health News, June 10, 2009. www.mobihealthnews.com.

Doll, Richard. 1955. "Mortality from Lung Cancer in Asbestos Workers." *British Journal of Industrial Medicine* 12:81–86.

Dorsey, E. Ray, Jason de Roulet, Joel P. Thompson, et al. 2010. "Funding of US Biomedical Research, 2003–2008." *Journal of the American Medical Association* 303:137–43.

Dorsey, E. Ray, Alexis Elbaz, Emma Nichols, et al. 2018. "Global, Regional, and National Burden of Parkinson's Disease, 1990–2016: A Systemic Analysis for the Global Burden of Disease Study 2016." *Lancet Neurology* 17:939–53.

Dorsey, E. Ray, Todd Sherer, Michael S. Okun, et al. 2018. "The Emerging Evidence of a Parkinson Pandemic." *Journal of Parkinson's Disease* 8:S3–S8.

Downes, Stephen M. 2019. "The Role of Ancient DNA Research in Archaeology." *Topoi* 40:285–93.

Dreifus, Claudia. 2008. "A Genetics Pioneer Sees a Bright Future, Cautiously." *New York Times,* April 29, 2008. www.nytimes.com.

———. 2009. "Researcher Behind the Drug Gleevec." *New York Times,* Nov. 2, 2009. www.nytimes.com.

Druker, Brian J., Moshe Talpaz, Debra J. Resta, et al. 2001. "Efficacy and Safety of a Specific Inhibitor of the BCR-ABL Tyrosine Kinase in Chronic Myeloid Leukemia." *New England Journal of Medicine* 344:1031–37.

Dugger, Sarah A., Adam Platt, and David Goldstein. 2018. "Drug Development in the Era of Precision Medicine." *Nature Reviews Drug Discovery* 17:183–96.

Duncan, Greg J., Nancy J. Kirkendall, and Constance F. Citro. 2014. *The National Children's Study: An Assessment.* Washington, D.C.: National Academies Press.

Dunlap, Riley E., and Angela G. Mertig, eds. 1992. *American Environmentalism: The U.S. Environmental Movement, 1970–1990.* Washington, D.C.: Taylor & Francis.

Dusetzina, Stacie B., Aaron N. Winn, Gregory A. Abel, et al. 2014. "Cost Sharing and Adherence to Tyrosine Kinase Inhibitors for Patients with Chronic Myeloid Leukemia." *Journal of Clinical Oncology* 32:306–11.

Dwyer, Terrence, Anne-Louise B. Ponsonby, Neville M. Newman, et al. 1991. "Prospective Cohort Study of Prone Sleeping Position and Sudden Infant Death Syndrome." *Lancet* 337:1244–47.

Edge. 2016. "The Genomic Ancient DNA Revolution: A New Way to Investigate the Past—a Conversation with David Reich." Feb. 1, 2016. www.edge.org.

Eichelbaum, Michel, Magnus Ingelman-Sundberg, and William E. Evans. 2006. "Pharmacogenomics and Individualized Drug Therapy." *Annual Review of Medicine* 57:119–37.

El-Husseini, Zaid W., Reinoud Gosens, Frank Dekker, et al. 2020. "The Genetics of Asthma and the Promise of Genomics-Guided Drug Target Therapy." *Lancet Respiratory Medicine* 8:1045–56.

Engleberts, A. C., and G. A. de Jonge. 1990. "Choice of Sleeping Position for Infants: Possible Association with Cot Death." *Archives of Disease in Childhood* 65:462–67.

Evans, Jennifer. 2010. "Francis Collins: DNA May Be a Doctor's Best Friend." National Public Radio, April 5, 2010. www.npr.org.

Executive Order 12898. 1994. "Federal Actions to Address Environmental Justice in Minority Populations and Low-Income Populations." Feb. 11, 1994. www.archives.gov.

Executive Order 13045. 1997. "Protection of Children from Environmental Health Risks and Safety Risks." April 21, 1997. www.archives.gov.

Experts in Chronic Myelogenous Leukemia. 2013. "The Price of Drugs for Chronic Myelogenous Leukemia (CML) Is a Reflection of the Unsustainable Prices of Cancer Drugs: From the Perspective of a Large Group of CML Experts." *Blood* 121:4439–42.

Eyal, Gil, Maya Sabatello, Kathryn Tabb, et al. 2019. "The Physician-Patient Relationship in the Age of Precision Medicine." *Genetics in Medicine* 21:813–15.

Falconi, Marta. 2014. "Novartis Manages to Push Back Competition to Leukemia Drug in the U.S." *Wall Street Journal,* May 15, 2014. www.wsj.com.

Farooq, Faheem, Peter J. Mogayzel, Sophie Lanzkron, et al. 2020. "Comparison of US Federal and Foundation Funding of Research for Sickle Cell Disease and Cystic Fibrosis and Factors Associated with Research Productivity." *JAMA Network Open* 3:e201737.

Farooq, Faheem, and John J. Strouse. 2018. "Disparities in Foundation and Federal Support and Development of New Therapeutics for Sickle Cell Disease and Cystic Fibrosis." *Blood* 132:4687.

Fee, Margery. 2006. "Racializing Narratives: Obesity, Diabetes, and the 'Aboriginal' Thrifty Genotype." *Social Science and Medicine* 62:2988–97.

Feldman, Eric A. 2011. "The Genetic Information Nondiscrimination Act (GINA): Public Policy and Medical Practice in the Age of Personalized Medicine." *Journal of General Internal Medicine* 27:743–46.

Feldman, Simon D., and Alfred I. Tauber. 1997. "Sickle Cell Anemia: Reexamining the First 'Molecular Disease.'" *Bulletin of the History of Medicine* 71:623–50.

Fernandes, Brisa S., Leanne M. Williams, Johann Steiner, et al. 2017. "The New Field of 'Precision Psychiatry.'" *BMC Medicine* 15:80.

Ferreira, Mariana Leal, and Gretchen Chesley Lang, eds. 2006. *Indigenous Peoples and Diabetes: Community Empowerment and Wellness.* Durham, N.C.: Carolina Academic Press.

Filipp, Fabian V. 2018. "Precision Dermatology—a Vibrant Field Where Diversity, Immunotherapy, and Precision Medicine Intersect." *BMC on Medicine,* Oct. 3, 2018. blogs.biomedcentral.com.

Fischer, Katrina A., Robert E. Miller, and John A. Glaspy. 2020. "Financial Toxicity from Generic Specialty Drug Use." *JAMA Oncology* 7:175–76.

Fleming, Peter J., Ruth Gilbert, Yehu Azaz, et al. 1990. "Interaction Between Bedding and Sleeping Position in the Sudden Infant Death Syndrome: A Population Based Case-Control Study." *British Medical Journal* 301:85–89.

Flippen, J. Brooks. 2000. *Nixon and the Environment.* Albuquerque: University of New Mexico Press.

Fox, Keolu. 2020. "The Illusion of Inclusion—The 'All of Us' Research Program and Indigenous Peoples' DNA." *The New England Journal of Medicine* 383:411–13.

Fox Keller, Evelyn. 2002. *The Century of the Gene.* Cambridge, Mass.: Harvard University Press.

Franklin, Benjamin. 1735. "On Protection of Towns from Fire." *Pennsylvania Gazette,* Feb. 4, 1735.

Franks, Paul W., Helen C. Looker, Sayuko Kobes, et al. 2006. "Gestational Glucose Tolerance and Risk of Type 2 Diabetes in Young Pima Indian Offspring." *Diabetes* 55:460–65.

Fredrickson, Leif, Christopher Sellers, Lindsey Dillon, et al. 2018. "History of US Presidential Assaults on Modern Environmental Health Protection." *American Journal of Public Health* 108:S95–S103.

Freudenberg, Nicholas. 2021. *At What Cost: Modern Capitalism and the Future of Health.* New York: Oxford University Press.

Fuchsberger, Christian, Jason Flannick, Tanya M. Teslovich, et al. 2016. "The Architecture of Type 2 Diabetes." *Nature* 536:41–47.

Fujimura, Joan. 1998. "The Molecular Biological Bandwagon in Cancer Research: Where Social Worlds Meet." *Social Problems* 35:261–83.

Fuller, Jonathan. 2017. "The New Medical Model: A Renewed Challenge for Biomedicine." *CMAJ* 189:E640–E641.

Furholt, Martin. 2018. "Massive Migrations? The Impact of Recent aDNA Studies on Our View of Third Millennium Europe." *European Journal of Archaeology* 21:159–91.

Galeon, Dom. 2016. "DNA-Based Curriculum? Genetics Could Dictate How Students Learn in the Future." Futurism, Nov. 30, 2016. futurism.com.

Galeone, Carla, Chiara Scelfo, Francesca Bertolini, et al. 2018. "Precision Medicine in Targeted Therapies for Severe Asthma: Is There Any Place for 'Omic' Technology?" *BioMed Research International,* article ID 4617565.

Garrod, Alfred Baring. 1909. *Inborn Errors of Metabolism.* Oxford: Oxford University Press.

Gasser, Thomas. 2015. "Usefulness of Genetic Testing in PD and PD Trials: A Balanced Review." *Journal of Parkinson's Disease* 5:209–15.

Gatto, Nicole M., Myles Cockburn, Jeff Bronstein, et al. 2009. "Well-Water Consumption and Parkinson's Disease in Rural California." *Environmental Health Perspectives* 117:1912–18.

Gaysina, Darya. 2016. "How Genetics Could Help Future Learners Unlock Hidden Potential." *Conversation,* Nov. 15, 2016. theconversation.com.

Geiger, Roger. 2004. *Knowledge and Money.* Stanford, Calif.: Stanford University Press.

Genetic Support Foundation. 2018. "Genetics and Personalized Medicine." www.geneticsupport.org.

GenomeWeb. 2008. "Personalized Medicine Policy Advocates Have High Hopes for New President." Nov. 7, 2008. www.genomeweb.com.

GenomeWeb News. 2007. "Illumina Closes Solexa Acquisition." Jan. 26, 2007. www.genomeweb.com.

Giangrande, Paul L. F. 2000. "The History of Blood Transfusion." *British Journal of Haematology* 110:758–67.

Gibbons, Ann. 2011. "Who Were the Denisovans?" *Science* 333:1084–87.

———. 2015. "Cave Was Lasting Home to Denisovans." *Science* 349:1270–71.

Gilbert, Ruth, Georgia Salanti, Melissa Harden, and Sarah See. 2005. "Infant Sleeping Position and the Sudden Infant Death Syndrome: Systematic Review of Observational Studies and Historical Review of Recommendations from 1940 to 2002." *International Journal of Epidemiology* 34:874–87.

Gilligan, Adrienne M., David S. Alberts, Denise J. Roe, et al. 2018. "Death or Debt? National Estimates of Financial Toxicity in Persons with Newly-Diagnosed Cancer." *American Journal of Medicine* 131:1187–99.

Ginsburg, Geoffrey S., and Kathryn A. Phillips. 2018. "Precision Medicine: From Science to Value." *Health Affairs* 37:694–701.

Gladwell, Malcolm. 1998. "The Pima Paradox." *New Yorker,* Feb. 2, 1998, 44–57.

Glennan, Stuart. 2017. *The New Mechanical Philosophy*. Oxford: Oxford University Press.

Glorikian, Harry. 2014. "Precision Medicine: New Opportunity in the Age of Cost-Conscious Medicine." *Genetic Engineering and Biotechnology News,* April 24, 2014. www.genengnews.com.

Goetz, Thomas. 2007. "23andMe Will Decode Your DNA for $1,000. Welcome to the Age of Genomics." *Wired,* Nov. 17, 2007. www.wired.com.

Gold, Jenny. 2017. "Sickle Cell Patients Fight Uphill Battle for Research, Treatment—and Compassion." *STAT,* Dec. 26, 2017. www.statnews.com.

Goldman, Samuel M. 2014. "Environmental Toxins and Parkinson's Disease." *Annual Review of Pharmacology and Toxicology* 54:141–64.

Goldman, Samuel M., Patricia J. Quinlan, G. Webster Ross, et al. 2011. "Solvent Exposures and Parkinson Disease Risk in Twins." *Annals of Neurology* 71:776–84.

Goldstein, David B., Sarah K. Tate, and Sanjay M. Sisodiya. 2003. "Pharmacogenetics Goes Genomic." *Nature Reviews Genetics* 4:937–47.

Gonzalez, Frank J., Radek C. Skodat, Shioko Kimura, et al. 1988. "Characterization of the Common Genetic Defect in Humans Deficient in Debrisoquine Metabolism." *Nature* 331:442.

Gormley, Melinda. 2007. "The First 'Molecular Disease': A Story of Linus Pauling, the Intellectual Patron." *Endeavour* 31:71–77.

Gosling, Anna L., Hallie R. Buckley, Elizabeth Matisoo-Smith, et al. 2015. "Pacific Populations, Metabolic Disease, and 'Just-So Stories': A Critique of the 'Thrifty Genotype' Hypothesis in Oceania." *Annals of Human Genetics* 79:470–80.

Gottlieb, Robert. 1993. *Forcing the Spring: The Transformation of the American Environmental Movement.* Washington, D.C.: Island Press.

Goulart, Bernardo H. L., Joseph M. Unger, Shasank Chennupati, et al. 2021. "Out-of-Pocket Costs for Tyrosine Kinase Inhibitors and Patient Outcomes in *EGFR*- and *ALK*-Positive Advanced Non-Small-Cell Lung Cancer." *JCO Oncology Practice* 17:e130–e139.

Grady, Denise. 1998. "Study Says Thousands Die from Reaction to Medicine." *New York Times,* April 15, 1998. www.nytimes.com.

Green, Richard E., Johannes Krause, Adrian W. Briggs, et al. 2010. "A Draft Sequence of the Neandertal Genome." *Science* 328:710–22.

Green, Sara, Annamarie Carusi, and Klaus Hoeyer. 2019. "Plastic Diagnostics: The Remaking of Disease and Evidence in Personalized Medicine." *Social Science and Medicine* 304. doi.org/10.1016/j.socscimed.2019.05.023.

Greenemeier, Larry. 2008. "Human Genome Project Head to Step Down." *Scientific American,* May 28, 2008. www.scientificamerican.com.

Greenwood, Veronique. 2011. "Why Are Asthma Rates Soaring?" *Scientific American,* April 1, 2011. www.scientificamerican.com.

Griffiths, Paul E., and Karola Stotz. 2013. *Genetics and Philosophy: An Introduction.* Cambridge, U.K.: Cambridge University Press.

Grigorenko, Elena. 2007. "How Can Genomics Inform Education?" *Mind, Brain, and Education* 1:20–27.

Grønnow, Bjarne. 1988. "Prehistory in Permafrost: Investigations at the Saqqaq Site, Qeqertasussuk, Disco Bay, West Greenland." *Journal of Danish Archaeology* 7:24–39.

———. 2012. "The Backbone of the Saqqaq Culture: A Study of the Nonmaterial Dimensions of the Early Arctic Small Tool Tradition." *Arctic Anthropology* 49:58–71.

———. 2016. "Independence I and Saqqaq: The First Greenlanders." In *The Oxford Handbook of the Prehistoric Arctic,* edited by Max Friesen and Owen Mason, 713–36. Oxford: Oxford University Press.

Grosse, Scott D., Michael S. Schechter, Roshni Kulkarni, et al. 2009. "Models of Comprehensive Multidisciplinary Care for Individuals in the United States with Genetic Disorders." *Pediatrics* 123:407–12.

Guarnieri, Michael, and John R. Balmes. 2014. "Outdoor Air Pollution and Asthma." *Lancet* 383:1581–92.

Guttmacher, Alan E., Steven Hirschfeld, and Francis S. Collins. 2013. "The National Children's Study—a Proposed Plan." *New England Journal of Medicine* 369:1873–75.

Haak, Wolfgang, Iosif Lazaridis, Nick Patterson, et al. 2015. "Massive Migration from the Steppe Was a Source for Indo-European Languages in Europe." *Nature* 522: 207–11.

Hakenbeck, Susanne. 2008. "Migration in Archaeology: Are We Nearly There Yet?" *Archaeological Review from Cambridge* 23:9–26.

Hamburg, Margaret A., and Francis S. Collins. 2010. "The Path to Personalized Medicine." *New England Journal of Medicine* 363:301–4.

Hamill, Sam, and J. P. Seaton. 1998. *The Essential Chuang Tzu.* Boston: Shambhala.

Hammonds, Evelynn M., and Susan Reverby. 2019. "Towards a Historically Informed Analysis of Racial Health Disparities Since 1619." *American Journal of Public Health* 109:1348–49.

Hampton, Elaine, and Cynthia C. Ontiveros. 2019. *Copper Stain: ASARCO's Legacy in El Paso.* Norman: University of Oklahoma Press.

Hanley, Robert. 1986. "Radon in Houses Is Viewed as Wider Threat in 3 States." *New York Times,* March 10, 1986. www.nytimes.com.

Hanushek, Eric A., and Ludger Wößmann. 2006. "Does Educational Tracking Affect Performance and Inequality? Differences-in-Differences Evidence Across Countries." *Economic Journal* 116:C63–C76.

Harden, Kathryn Paige. 2021. *The Genetic Lottery: Why DNA Matters for Social Equality.* Princeton, N.J.: Princeton University Press.

Hardy, Janet B. 2003. "The Collaborative Perinatal Project: Lessons and Legacy." *Annals of Epidemiology* 13:303–11.

Hardy, Janet B., Joseph S. Drage, and Esther C. Jackson. 1979. *The First Year of Life: The Collaborative Perinatal Project of the National Institute of Neurological and Communicative Disorders and Stroke.* Baltimore: Johns Hopkins University Press.

Harmon, Katherine. 2012. "New DNA Analysis Shows Ancient Humans Interbred with Denisovans." *Nature,* Aug. 31, 2012. www.nature.com.

Harold, John Gordon. 2018. "The Evolution of Personalized Medicine." *Cardiology Magazine,* Oct. 19, 2018. www.acc.org.

Harris, Gardiner. 2009. "Pick to Lead Health Agency Draws Praise and Some Concern." *New York Times,* July 8, 2009. www.nytimes.com.

Harris, Robert Z., Graham R. Jang, and Shirley Tsunoda. 2003. "Dietary Effects on Drug Metabolism and Transport." *Clinical Pharmacokinetics* 42:1071–88.

Hart, Sara. 2016. "Precision Education Initiative: Moving Toward Personalized Education." *Mind, Brain, and Education* 10:209–11.

Harvard Business Review Analytic Services. 2018. "Expanding Precision Medicine: The Path to Higher Value." White paper. www.siemens-healthineers.com.

Hathaway, William. 1997. "Decoding Dr. Ruaño." *Hartford Courant,* Dec. 28, 1997. www.courant.com.

Havel, Erin. 2012. *The Malformation of Health Care: Illness, Insurance, and a Young American's Struggle for Survival.*

———. 2015. "A Generic Cancer Drug Is Coming to America . . . Great." Stupid Cancer Stories, Dec. 4, 2015. blog.stupidcancer.org.

Hawks, John. 2021. "Accurate Depiction of Uncertainty in Ancient DNA Research: The Case of Neandertal Ancestry in Africa." *Journal of Social Archaeology,* Feb. 24, 2021.

Hedgecoe, Adam. 2003. "Terminology and the Construction of Scientific Disciplines: The Case of Pharmacogenomics." *Science, Technology, and Human Values* 28:513–37.

———. 2004. *The Politics of Personalised Medicine: Pharmacogenetics in the Clinic.* Cambridge, U.K.: Cambridge University Press.

Hedgecoe, Adam, and Paul Martin. 2003. "The Drugs Don't Work: Expectations and the Shaping of Pharmacogenetics." *Social Studies of Science* 33:327–64.

Hegner, Richard E. 2000. "The Asthma Epidemic: Prospects for Controlling an Escalating Public Health Crisis." *National Health Policy Forum,* Sept. 2000, 1–16.

Henderson, Rebecca. 2007. "Eli Lilly's Project Resilience: Anticipating the Future of the Pharmaceutical Industry." Massachusetts Institute of Technology (MIT) Case. mitsloan.mit.edu.

———. 2008. "Eli Lilly: Recreating Drug Discovery for the 21st Century." With Cate Reavis. Massachusetts Institute of Technology (MIT) Case. mitsloan.mit.edu.

Henk, Henry J., Lena E. Winestone, Jennifer J. Wilkes, et al. 2020. "Trend in TKI Use, Adherence, and Switching Patterns in Patients with CML: Before and After the Availability of Generic Imatinib." *Journal of Clinical Pathways* 6:35–42.

Herbst, Roy S., Daniel Morgensztern, and Chris Boshoff. 2018. "The Biology and Management of Non-Small Cell Lung Cancer." *Nature* 553:446–54.

Hernandez-Pacheco, Natalia, Maria Pino-Yanes, and Carlos Flores. 2019. "Genomic Predictors of Asthma Phenotypes and Treatment Response." *Frontier in Pediatrics* 7:6.

Heyd, Volker. 2017. "Kossinna's Smile." *Antiquity* 91:348–59.

Hicken, Margaret T., Gilbert C. Gee, Jeffrey Morenoff, et al. 2012. "A Novel Look at Racial Health Disparities: The Interaction Between Social Disadvantage and Environmental Health." *American Journal of Public Health* 102:2344–51.

Hiltzik, Michael. 2014. "How a $1.3-Billion, 21-Year Study of U.S. Children's Health Fell to Pieces." *Los Angeles Times*, December 19, 2014. latimes.com.

Hippocrates. 1923. *Prognostic. Regimen in Acute Diseases. The Sacred Disease. The Art. Breaths. Law. Decorum. Physician (Ch. 1). Dentition.* Translated by W. H. S. Jones. Cambridge, Mass.: Harvard University Press.

———. 1968. *On Wounds in the Head. In the Surgery. Fractures, Joints, Mochlicon.* Vol. 3. Translated by E. T. Withington. Cambridge, Mass.: Harvard University Press.

———. 1988. *Diseases 3. Internal Affections. Regimen in Acute Diseases.* Vol. 6. Translated by Paul Potter. Cambridge, Mass.: Harvard University Press.

Hoberman, John. 2012. *Black and Blue: The Origin and Consequences of Medical Racism.* Berkeley: University of California Press.

Hockwald, Capt. Robert S., Major John Arnold, Capt. Charles B. Clayman, et al. 1952. "Toxicity of Primaquine in Negroes." *Journal of the American Medical Association* 149:1568–70.

Holden, Arthur L. 2002. "The SNP Consortium: Summary of a Private Consortium Effort to Develop an Applied Map of the Human Genome." *BioTechniques* 32:S22–S26.

Holden, Constance. 2009. "Collins Dominates Rumor Mill for NIH Director." *Science*, Jan. 2, 2009. www.sciencemag.org.

Hopkins, Michael M., Dolores Ibarreta, Sibylle Gaisser, et al. 2006. "Putting Pharmacogenomics into Practice." *Nature Biotechnology* 24:403–10.

Hornblum, Allen M. 1997. "They Were Cheap and Available: Prisoners as Research Subjects in Twentieth Century America." *British Medical Journal* 315:1437–41.

Horowitz, Milton. 1962. "Trends in Education." *Journal of Medical Education* 37:634–37.

Horsburgh, K. Ann. 2015. "Molecular Anthropology: The Judicial Use of Genetic Data in Archaeology." *Journal of Archaeological Science* 56:141–45.

Housman, David, and Fred D. Ledley. 1998. "Why Pharmacogenomics? Why Now?" *Nature Biotechnology* 16:492–93.

Howard, Alan, Marcelo A. Fernandez-Vina, Frederick R. Appelbaum, et al. 2015. "Recommendations for Donor HLA Assessment and Matching for Allogeneic Stem Cell Transplantation: Consensus Statement of the Blood and Marrow Transplant Clinical Trials Network (BMT CTN)." *Biology of Blood and Marrow Transplantation* 21:4–7.

Hrdlička, Aleš. 1908. *Physiological and Medical Observations Among the Indians of the Southwestern United States and Northern Mexico.* Washington, D.C.: Government Printing Office.

Hudson, Maui, Nanibaa' A. Garrison, Rogena Sterling, et al. 2020. "Rights, Interests and Expectations: Indigenous Perspectives on Unrestricted Access to Genomic Data." *Nature Reviews Genetics* 21:377–384.

Huntington, Scott F., Amy J. Davidoff, and Cary P. Gross. 2019. "Precision Medicine in Oncology II: Economics of Targeted Agents and Immuno-oncology Drugs." *Journal of Clinical Oncology* 38:351–59.

Huntington's Disease Collaborative Research Group. 1993. "A Novel Gene Containing a Trinucleotide Repeat That Is Expanded and Unstable on Huntington's Disease Chromosomes." *Cell* 72:971–83.

Hynes, H. Patricia. 1985. "Ellen Swallow, Lois Gibbs, and Rachel Carson: Catalysts of the American Environmental Movement." *Women's Studies International Forum* 8:291–98.

Idle, Jeffrey R., and Robert L. Smith. 1995. "Pharmacogenetics in the New Patterns of Healthcare Delivery." *Pharmacogenetics* 5:347–50.

"Interview with Senator Barack Obama About Genomics and Personalized Medicine Act of 2006." 2007. *Clinical Advances in Hematology and Oncology* 5:39–40.

Jackson Laboratory. n.d. "What Is Personalized Medicine?" www.jax.org.

Jacques, Peter J., Riley E. Dunlap, and Mark Freeman. 2008. "The Organisation of Denial: Conservative Think Tanks and Environmental Scepticism." *Environmental Policy* 17:349–85.

Jain, Nilesh. 2019. "How Precision Medicine Will Change the Future of Healthcare." World Economic Forum, Jan. 1, 2019. www.weforum.org.

Joe, Jennie R., and Robert S. Young, eds. 1994. *Diabetes as a Disease of Civilization: The Impact of Cultural Change on Indigenous Peoples.* Berlin: Mouton de Gruyter.

Johnson, Amber L., ed. 2004. *Processual Archaeology: Exploring Analytical Strategies, Frames of Reference, and Culture Process.* Westport, Conn.: Praeger.

Johnson, Carolyn Y. 2016. "This Drug Is Defying a Rare Form of Leukemia—and It Keeps Getting Pricier." *Washington Post,* March 9, 2016. www.washingtonpost.com.

Jones, David S. 2005. "The Persistence of American Indian Health Disparities." *American Journal of Public Health* 96:2122–34.

———. 2013. "How Personalized Medicine Became Genetic, and Racial: Werner Kalow and the Formations of Pharmacogenetics." *Journal of the History of Medicine and Allied Sciences* 68:1–48.

Jones, David S., and Gerald M. Oppenheimer. 2017. "If the Framingham Heart Study Did Not Invent the Risk Factor, Who Did?" *Perspectives in Biology and Medicine* 60: 131–50.

Jones, Elizabeth. 2019. "Ancient Genetics to Ancient Genomics: Celebrity and Credibility in Data-Driven Practice." *Biology and Philosophy* 34:27.

Jones, Elizabeth D., and Elsbeeth Bösl. 2021. "Ancient Human DNA: A History of Hype (Then and Now)." *Journal of Social Archaeology,* Feb. 18, 2021.

Jones, Mark. 2014. "Larry Bock Interview Conducted by Mark Jones." San Diego Technology Archive, June 10, 2014. library.ucsd.edu.

Jones, Steven. 2006. *The Ghost Map: The Story of London's Most Terrifying Epidemic—*

and How It Changed Science, Cities, and the Modern World. New York: Riverhead Books.

Joseph, Charles, and Adam Joel Mrdjenovich. 2017. "Personalized Medicine, Genomics, and Enhancement: Monuments to Neoliberalism." *American Journal of Clinical and Experimental Medicine* 5:75–92.

Joyce, Christopher. 2010. "2010: A Good Year for Neanderthals (and DNA)." National Public Radio, Dec. 28, 2010. www.npr.org.

Joyner, Michael J. 2015. "'Moonshot' Medicine Will Let Us Down." *New York Times,* Jan. 29, 2015. www.nytimes.com.

Joyner, Michael J., and Nigel Paneth. 2015. "Seven Questions for Precision Medicine." *Journal of the American Medical Association* 314:999–1000.

———. 2019a. "Promises, Promises, and Precision Medicine." *Journal of Clinical Investigation* 129:946–48.

———. 2019b. "Precision Medicine's Rosy Predictions Haven't Come True. We Need Fewer Promises and More Debate." *STAT,* Feb. 7, 2019. www.statnews.com.

Juengst, Eric, Michelle L. McGowan, Jennifer R. Fishman, et al. 2016. "From 'Personalized' Medicine to 'Precision' Medicine: The Ethical and Social Implications of Rhetorical Reform in Genomic Medicine." *Hastings Center Report* 46:21–33.

Kaiser, Jocelyn. 2009a. "More Turmoil over Hidden Costs of NIH Children's Study." *Science,* Aug. 10, 2009. www.sciencemag.org.

———. 2009b. "Francis Collins: Looking Beyond the Funding Deluge." *Science,* Oct. 9, 2009.

———. 2012. "Two Advisors Now Gone from National Children's Study." *Science,* March 19, 2012. www.sciencemag.org.

———. 2015. "Blueprint in Hand, NIH Embarks on Study of a Million People." *Science,* Sept. 17, 2015. www.sciencemag.org.

———. 2018. "A Cancer Drug Tailored to Your Tumor? Experts Trade Barbs over 'Precision Oncology.'" *Science,* April 24, 2018. www.sciencemag.org.

Kaiser Health News. 2010. "NIH Chief Francis Collins: Medical Research 'Ought to Tell Us What Works.'" April 6, 2010. khn.org.

Källén, Anna, Charlotte Mulcare, Andreas Nyblom, et al. 2021. "Introduction: Transcending the aDNA Revolution." *Journal of Social Archaeology,* Feb. 20, 2021.

Kalow, Werner. 1962. *Pharmacogenetics: Heredity and the Response to Drugs.* Philadelphia: W. B. Saunders.

Kaplan, Abraham. 1964a. "The Age of the Symbol: A Philosophy of Library Education." *Library Quarterly* 34:295–304.

———. 1964b. *The Conduct of Inquiry: Methodology for Behavioral Science.* San Francisco: Chandler.

Kaplan, Sheila. 2016. "The NIH, in Pursuit of Precision Medicine, Tries to Avoid Ghosts of Its Past." *STAT,* Jan. 11, 2016. www.statnews.com.

Karow, Julia. 2001. "Reading the Book of Life." *Scientific American,* Feb. 12, 2001. www.scientificamerican.com.

Katsnelson, Alla. 2013. "Momentum Grows to Make 'Personalized' Medicine More 'Precise.'" *Nature Medicine* 19:249.

Keet, Corinne A., Meredith C. McCormack, Craig E. Pollack, et al. 2015. "Neighborhood

Poverty, Urban Residence, Race/Ethnicity, and Asthma: Rethinking the Inner-City Asthma Epidemic." *Journal of Allergy and Clinical Immunology* 135:655–62.

Kehl, Kenneth L., Christopher S. Lathan, Bruce E. Johnson, et al. 2019. "Race, Poverty, and Initial Implementation of Precision Medicine for Lung Cancer." *JNCI: Journal of the National Cancer Institute* 111:431–34.

Kevles, Daniel. 1985. *In the Name of Eugenics: Genetics and the Uses of Human Heredity.* Berkeley: University of California Press.

Khoury, Muin. 2016. "The Shift from Personalized Medicine to Precision Medicine and Precision Public Health: Words Matter!" blogs.cdc.gov.

Kindela, David. 2013. *DDT and the American Century: Global Health, Environmental Politics, and the Pesticide That Changed the World.* Chapel Hill: University of North Carolina Press.

Klass, Perri. 2020. *A Good Time to Be Born: How Science and Public Health Gave Children a Future.* New York: W. W. Norton.

Klebanoff, Mark A. 2009. "The Collaborative Perinatal Project: A 50-Year Retrospective." *Paediatric and Perinatal Epidemiology* 23:2–8.

Kline, Wendy. 2005. *Building a Better Race: Gender, Sexuality, and Eugenics from the Turn of the Century to the Baby Boom.* Berkeley: University of California Press.

Knipper, Corina, Alissa Mittnik, Ken Massy, et al. 2017. "Female Exogamy and Gene Pool Diversification at the Transition from the Final Neolithic to the Early Bronze Age in Central Europe." *Proceedings of the National Academy of Sciences* 114:10083–88.

Knowler, William C., David J. Pettitt, Peter H. Bennett, et al. 1983. "Diabetes Mellitus in the Pima Indians: Genetic and Evolutionary Considerations." *American Journal of Physical Anthropology* 62:107–14.

Knowler, William C., David J. Pettitt, Mohammed F. Saad, et al. 1990. "Diabetes Mellitus in the Pima Indians: Incidence, Risk Factors, and Pathogenesis." *Diabetes/Metabolism Research and Reviews* 6:1–27.

Knox, Robert. 2012. "Duxbury Celebrates Rachel Carson's 'Silent Spring.'" *Boston Globe,* May 24, 2012. www.bostonglobe.com.

Kohler, Robert E. 1994. *Lords of the Fly:* Drosophila *Genetics and the Experimental Life.* Chicago: University of Chicago Press.

Kolata, Gina. 1999. "Using Gene Tests to Customize Medical Treatment." *New York Times,* Dec. 20, 1999. www.nytimes.com.

Kolb, Hubert, and Stephan Martin. 2017. "Environmental/Lifestyle Factors in the Pathogenesis and Prevention of Type 2 Diabetes." *BMC Medicine* 15:131.

Konishi, Ikuo, ed. 2017. *Precision Medicine in Gynecology and Obstetrics.* Singapore: Springer Singapore.

Konstantinidou, Meropi K., Makrina Karaglani, Maria Panagopoulou, et al. 2017. "Are the Origins of Precision Medicine Found in the *Corpus Hippocraticum*?" *Molecular Diagnosis and Therapy* 21:601–6.

Kovas, Yulia, Sergey Malykh, and Darya Gaysina, eds. 2016. *Behavioural Genetics for Education.* London: Palgrave Macmillan.

Krause, Johannes, Qiaomei Fu, Jeffrey M. Good, et al. 2010. "The Complete Mitochondrial DNA Genome of an Unknown Hominin from Southern Siberia." *Nature* 464:894–97.

Krieger, Nancy. 2011. *Epidemiology and the People's Health: Theory and Context.* New York: Oxford University Press.

Kuderer, Nicole M., Kimberly A. Burton, Sibel Blau, et al. 2017. "Comparison of Two Commercially Available Next-Generation Sequencing Platforms in Oncology." *JAMA Oncology* 3:996–98.

Kuska, Bob. 1998. "Beer, Bethesda, and Biology: How 'Genomics' Came into Being." *Journal of the National Cancer Institute* 90:93.

Lambert, Jonathan. 2019. "Huge US Government Study to Offer Genetic Counselling." *Nature* 572:573.

Landrigan, Philip J., Robert W. Baloh, William F. Barthel, et al. 1975. "Neurophysiological Dysfunction in Children with Chronic Low-Level Lead Absorption." *Lancet* 305:708–12.

Landrigan, Philip J., Stephen H. Gehlbach, Bernard Rosenblum, et al. 1975. "Epidemic Lead Absorption Near an Ore Smelter—the Role of Particulate Lead." *New England Journal of Medicine* 292:123–29.

Landrigan, Philip J., Leonardo Trasande, Lorna E. Thorpe, et al. 2006. "The National Children's Study: A 21-Year Prospective Study of 100,000 American Children." *Pediatrics* 118:2173–86.

Landry, Latrice G., Nadya Ali, David R. Williams, et al. 2018. "Lack of Diversity in Genomic Databases Is a Barrier to Translating Precision Medicine Research into Practice." *Health Affairs* 37. www.healthaffairs.org.

Langreth, Robert, and Michael Waldholz. 1999. "Genetic Mapping Ushers in New Era of Profitable Personalized Medicines." *Wall Street Journal,* April 16, 1999. www.wsj.com.

Langreth, Robert, Michael Waldholz, and Stephen D. Moore. 1999. "Drug Firms Discuss Linking Up to Pursue Disease-Causing Genes." *Wall Street Journal,* March 4, 1999. www.wsj.com.

Lantz, Paula M., David Mendez, and Martin A. Philbert. 2013. "Radon, Smoking, and Lung Cancer: The Need to Refocus Radon Control Policy." *American Journal of Public Health* 103:443–47.

Lanzkron, Sophie. 2018. "Race, Funding, and Access to High Quality Care." *Closler,* April 4, 2018. closler.org.

Largent, Mark A. 2011. *Breeding Contempt: The History of Coerced Sterilization in the United States.* New Brunswick, N.J.: Rutgers University Press.

Lash, Jonathan, Katherine Gillman, and David Sheridan. 1984. *A Season of Spoils: The Reagan Administration's Attack on the Environment.* New York: Pantheon Books.

Lauerman, John. 2008. "Obama Should Tap Personalized Medicine Tools, Leavitt Says." *Deseret News,* Nov. 15, 2008. www.deseret.com.

Layzer, Judith A. 2012. *Open for Business: Conservatives' Opposition to Environmental Regulation.* Cambridge, Mass.: MIT Press.

Lazarou, Jason, Bruce H. Pomeranz, and Paul N. Corey. 1998. "Incidence of Adverse Drug Reactions in Hospitalized Patients." *Journal of the American Medical Association* 279: 1200–1205.

Lear, Linda. 1997. *Rachel Carson: Witness for Nature.* New York: Henry Holt.

Leavitt, Michael, and Gregory O. Downing. 2008. "Toward a Future of Personalized Cancer Care." *Cancer* (Supplement 113): 1724–27.

Lebl, Michal. 2003. "Centrifugation Based Automated Synthesis Technologies." *JALA: Journal of the Association for Laboratory Automation* 8:30–35.

Lee, Sandra Soo-Jin. 2009. "Pharmacogenomics and the Challenge of Health Disparities." *Public Health Genomics* 12:170–79.

Lee, Sandra Soo-Jin, and Ashwin Mudaliar. 2009. "Racing Forward: The Genomics and Personalized Medicine Act." *Science* 323:342.

Leigh, Vivian. 2020. "'We Need Answers, We Need Them Quickly': Veterans Push for Federal Cancer Study." WUSA9, Aug. 19, 2020. www.wusa9.com.

Lemoine, Mäel. 2017. "Neither from Words, nor from Visions: Understanding P-Medicine from Innovative Treatments." *Lato Sensu* 4:12–23.

Levy, Daniel, and Susan Brink. 2005. *A Change of Heart: How the People of Framingham, Massachusetts, Helped Unravel the Mysteries of Cardiovascular Disease.* New York: Alfred A. Knopf.

Lewis-Kraus, Gideon. 2019. "Is Ancient DNA Research Revealing New Truths—or Falling into Old Traps?" *New York Times Magazine,* Jan. 17, 2019. www.nytimes.com.

Li, Jie Jack. 2014. *Blockbuster Drugs: The Rise and Decline of the Pharmaceutical Industry.* Oxford: Oxford University Press.

Life. 1945. "Prison Malaria: Convicts Expose Themselves to Disease So Doctors Can Study It." June 4, 1945, 43–46.

Lipner, Ettie M., and David A. Greenberg. 2018. "The Rise and Fall of Linkage Analysis as a Technique for Finding and Characterizing Inherited Influences on Disease Expression." In *Disease Gene Identification: Methods and Protocols,* edited by Johanna K. DiStefano, 381–97. 2nd ed. New York: Humana Press.

Lombardino, Joseph G. 2000. "A Brief History of Pfizer Central Research." *Bulletin for the History of Chemistry* 25:10–15.

Lombardo, Paul A. 2008. *Three Generations, No Idiots: Eugenics, the Supreme Court, and Buck v. Bell.* Baltimore: Johns Hopkins University Press.

Loveless, Tom. 1999. *The Tracking Wars: State Reform Meets School Policy.* Washington, D.C.: Brookings Institution Press.

———. 2013. "The Resurgence of Ability Grouping and Persistence of Tracking." March 18, 2013. www.brookings.edu.

Löwy, Ilana. 2011. "Historiography of Biomedicine: 'Bio,' 'Medicine,' and in Between." *Isis* 102:116–22.

Lyman, Gary H., and Henry J. Henk. 2020. "Association of Generic Imatinib Availability and Pricing with Trends in Tyrosine Kinase Inhibitor Use in Patients with Chronic Myelogenous Leukemia." *JAMA Oncology* 6:1969–70.

Lyons, Michele. 2006. *70 Acres of Science: The National Institute of Health Moves to Bethesda.* Office of NIH History, National Institutes of Health. history.nih.gov.

Lytle, Mark Hamilton. 2007. *The Gentle Subversive: Rachel Carson, "Silent Spring," and the Rise of the Environmental Movement.* New York: Oxford University Press.

Maher, Brendan. 2008. "Personal Genomes: The Case of the Missing Heritability." *Nature* 456:18–21.

Mahmood, Syed S., Daniel Levy, Ramachandran S. Vasan, et al. 2013. "The Framingham Heart Study and the Epidemiology of Cardiovascular Disease: A Historical Perspective." *Lancet* 383:999–1008.

Mäki, Uskali, Adrian Walsh, and Manuela Fernández Pinto, eds. 2018. *Scientific Imperialism: Exploring the Boundaries of Interdisciplinarity.* New York: Routledge.

Mancinelli, Laviero, Maureen Cronin, and Wolfgang Sadée. 2000. "Pharmacogenomics: The Promise of Personalized Medicine." *AAPS PharmSci* 2:29–41.

Maner, Brent. 2018. *Germany's Ancient Past: Archaeological and Historical Interpretation Since 1700.* Chicago: University of Chicago Press.

Manolio, Teri A., Lisa D. Brooks, and Francis S. Collins. 2008. "A HapMap Harvest of Insights into the Genetics of Common Disease." *Journal of Clinical Investigation* 118:1590–605.

Manolio, Teri A., and Francis S. Collins. 2009. "The HapMap and Genome-Wide Association Studies in Diagnosis and Therapy." *Annual Review of Medicine* 60:443–56.

Manolio, Teri A., Francis S. Collins, Nancy J. Cox, et al. 2008. "Finding the Missing Heritability of Complex Diseases." *Nature* 461:747–53.

Mapes, Diane. 2015. "Cruel Choices: Buy Lifesaving Meds or Groceries?" Hutch News Stories, June 3, 2015. www.fredhutch.org.

———. 2020. "New Cancer Drugs Lead to Life-Threatening Financial Choices." Hutch News Stories, Dec. 7, 2020. www.fredhutch.org.

Marcon, Alessandro R., Mark Bieber, and Timothy Caulfield. 2018. "Representing a 'Revolution': How the Popular Press Has Portrayed Personalized Medicine." *Genetics in Medicine* 20:950–56.

Marcosson, Isaac Frederick. 1949. *Metal Magic: The Story of the American Smelting and Refining Company.* New York: Farrar, Straus.

Marcus, Amy Dockser. 2019. "The Unfulfilled Promise of DNA Testing." *Wall Street Journal,* May 17, 2019. www.wsj.com.

Markowitz, Gerald, and David Rosner. 2013. *Lead Wars: The Politics of Science and the Fate of America's Children.* Berkeley: University of California Press.

Marquart, John, Emerson Y. Chen, and Vinay Prasad. 2018. "Estimation of the Percentage of US Patients with Cancer Who Benefit from Genome-Driven Oncology." *JAMA Oncology* 4:1093–98.

Marras, C., J. C. Beck, J. H. Bower, et al. 2018. "Prevalence of Parkinson's Disease Across North America." *NPJ Parkinson's Disease* 4:21.

Marshall, Andrew. 1997a. "Laying the Foundations for Personalized Medicines." *Nature Biotechnology* 15:954–57.

———. 1997b. "Getting the Right Drug into the Right Patient." *Nature Biotechnology* 15:1249–52.

———. 1998. "One Drug Does Not Fit All." *Nature Biotechnology* 16 (Supplement): 1.

Marshall, Carrie, and James Stables. 2020. "The Story of Fitbit: How a Wooden Box Was Bought by Google for $2.1Bn." Wareable, April 4, 2020. www.wareable.com.

Marson, Fernando A. L., Carmen S. Bertuzzo, and Jose D. Ribeiro. 2017. "Personalized or Precision Medicine? The Example of Cystic Fibrosis." *Frontiers in Pharmacology* 8: Article 390.

Martschenko, Daphne. 2020a. "'The Train Has Left the Station': The Arrival of the Biosocial Sciences in Education." *Research in Education* 107:3–9.

———. 2020b. "'DNA Dreams': Teacher Perspectives on the Role and Relevance of Genetics for Education." *Research in Education* 107:33–54.

Martschenko, Daphne, and Lucas J. Matthews. 2020. "Genomics, Behavior, and Social Outcomes." Hastings Center Bioethics Briefings, Dec. 1, 2020. www.thehastingscenter.org.

Martschenko, Daphne, Sam Trejo, and Benjamin W. Domingue. 2019. "Genetics and Education: Recent Developments in the Context of an Ugly History and an Uncertain Future." *AERA Open* 5:1–15.

Maslow, Abraham. 1966. *The Psychology of Science: A Reconnaissance.* New York: Harper & Row.

Massard, Christophe, Stefan Michiels, Charles Ferté, et al. 2017. "High-Throughput Genomics and Clinical Outcomes in Hard-to-Treat Advanced Cancers: Results of the MOSCATO 01 Trial." *Cancer Discovery* 7:1–10.

"Matsui Urges Appropriators to Save the National Children's Study." May 11, 2006. matsui.house.gov.

Matthew, Dayna Bowen. 2015. *Just Health: A Cure for Racial Inequality in American Health Care.* New York: New York University Press.

Matthews, Lucas J., Matthew S. Lebowitz, Ruth Ottman, et al. 2021. "Pygmalion in the Genes? On the Potentially Negative Impacts of Polygenic Scores for Educational Attainment." *Social Psychology of Education* 24:789–808.

Matthews, Lucas J., and Eric Turkheimer. 2021. "Across the Great Divide: Pluralism and the Hunt for Missing Heritability." *Synthese* 198:2297–311. doi.org/10.1007/s11229-019-02205-w.

McCarthy, Mark I. 2017. "Painting a New Picture of Personalised Medicine for Diabetes." *Diabetologia* 60:793–99.

McCarty, Catherine A., Rex L. Chisholm, Christopher G. Chute, et al. 2011. "The eMERGE Network: A Consortium of Biorepositories Linked to Electronic Medical Records Data for Conducting Genomic Studies." *BMC Medical Genomics* 4:13.

McDermott, Robyn. 1998. "Ethics, Epidemiology, and the Thrifty Gene: Biological Determinism as a Health Hazard." *Social Science and Medicine* 47:1189–95.

McGlone, Kathleen, Erika Blacksher, and Wylie Burke. 2017. "Genomics, Health Disparities, and Missed Opportunities for the Nation's Research Agenda." *Journal of the American Medical Association* 317:1831–32.

McGurty, Eileen. 2007. *Transforming Environmentalism: Warren County, PCBs, and the Origins of Environmental Justice.* New Brunswick, N.J.: Rutgers University Press.

McNamee, Gregory. 2012. *Gila: The Life and Death of an American River.* Albuquerque: University of New Mexico Press.

McVean, Gil, Chris C. Spencer, and Raphaelle Chaix. 2005. "Perspectives on Human Genetic Variation from the HapMap Project." *PLOS Genetics* 1:e54.

Meek, James, and Michael Ellison. 2000. "On the Path of Biology's Holy Grail." *Guardian,* June 4, 2000. www.theguardian.com.

Mees, Bernard. 2008. *The Science of the Swastika.* Budapest: Central European University Press.

Meldgaard, Morten. 2004. *Ancient Harp Seal Hunters of Disko Bay: Subsistence and Settlement at the Saqqaq Culture Site Qeqertasussuk (2400–1400 BC), West Greenland.* Copenhagen: Museum Tusculanum Press.

Menke, Andy, Sarah Casagrande, Linda Geiss, et al. 2015. "Prevalence of and Trends in Diabetes Among Adults in the United States, 1988–2012." *JAMA* 314:1021–29.

Metspalu, Andres. 2004. "The Estonian Genome Project." *Drug Development Research* 62:97–101.

Meyer, Mathias, Martin Kircher, Marie-Theres Gansauge, et al. 2012. "A High-Coverage Genome Sequence from an Archaic Denisovan Individual," *Science* 338:222–26.

Meyers, Deborah A., Eugene R. Bleecker, John W. Holloway, et al. 2014. "Asthma Genetics and Personalised Medicine." *Lancet Respiratory Medicine* 2:405–15.

Michael, David. 2008. *Doubt Is Their Product: How Industry's Assault on Science Threatens Your Health.* Oxford: Oxford University Press.

Michael, Karri L., Laura C. Taylor, Sandra L. Schultz, et al. 1998. "Randomly Ordered Addressable High-Density Optical Sensor Arrays." *Analytical Chemistry* 70:1242–48.

Michael, Robert T., and Colm A. O'Muircheartaigh. 2008. "Design Priorities and Disciplinary Perspectives: The Case of the U.S. National Children's Study." *Journal of the Royal Statistical Society* 171:465–80.

Michl, Susanne. 2015. "Inventing Traditions, Raising Expectations. Recent Debates on 'Personalized Medicine.'" In *Individualized Medicine: Ethical, Economic, and Historical Perspectives,* edited by Tobias Fischer, Martin Langanke, Paul Marschall, and Susanne Michl, 45–60. Heidelberg: Springer.

Millar, Heather. 2015. "A Different World: Metastatic Breast Cancer." CURE: Cancer Updates, Research, and Education, Oct. 23, 2015. www.curetoday.com.

Miller, Franklin G. 2013. "The Stateville Penitentiary Malaria Experiments: A Case Study in Retrospective Ethical Assessment." *Perspectives in Biology and Medicine* 56:548–67.

Miller, Max, Thomas A. Burch, Peter H. Bennett, et al. 1965. "Prevalence of Diabetes Mellitus in the American Indians: Results of Glucose Tolerance Tests in the Pima Indians of Arizona." *Diabetes* 14:439–40.

Mills, Kay. 1998. "Duane Alexander." *Los Angeles Times,* Oct. 11, 1998. www.latimes.com.

Miners, John O., and Donald J. Birkett. 1998. "Cytochrome P4502C9: An Enzyme of Major Importance in Human Drug Metabolism." *British Journal of Clinical Pharmacology* 45:525–38.

Mirowski, Philip. 2011. *Science-Mart: Privatizing American Science.* Cambridge, Mass.: Harvard University Press.

Mirowski, Philip, and Esther-Mirjam Sent, eds. 2002. *Science Bought and Sold.* Chicago: University of Chicago Press.

Monmaney, Terence. 2011. "A Triumph in the War Against Cancer." *Smithsonian Magazine,* May 2011. www.smithsonianmag.com.

Montoya, Michael J. 2011. *Making the Mexican Diabetic: Race, Science, and the Genetics of Inequality.* Berkeley: University of California Press.

Montrie, Chad. 2018. *The Myth of "Silent Spring": Rethinking the Origins of American Environmentalism.* Oakland: University of California Press.

Morbidity and Mortality Weekly Report. 1973. "Human Lead Absorption—Texas." 22:405–7. www.cdc.gov.

———. 1996. "Sudden Infant Death Syndrome—United States, 1983–1994." 40:859–63. www.cdc.gov.

Morin, Bode J. 2010. *The Legacy of American Copper Smelting: Industrial Heritage Versus Environmental Policy.* Knoxville: University of Tennessee Press.

Morrison, Chris. 2015. "'Financial Toxicity' Looms as Cancer Combinations Proliferate." *Nature Biotechnology* 33:783–84.

Morse, Susan. 2019. "Precision vs. Personalized Medicine: What's the Difference?" Healthcare Finance, May 21, 2019. www.healthcarefinancenews.com.

Moses, Hamilton, III, E. Ray Dorsey, David H. M. Matheson, et al. 2005. "Financial Anatomy of Biomedical Research." *Journal of the American Medical Association* 294:1333–42.

Motulsky, Arno G. 1957. "Drug Reactions, Enzymes, and Biochemical Genetics." *Journal of the American Medical Association* 165:835–37.

Motulsky, Arno G., as told to Mary-Claire King. 2016. "The Great Adventure of an American Human Geneticist." *Annual Review of Genomics and Human Genetics* 17:1–15.

Motulsky, Harvey, and Gail P. Jarvik. 2018. "A German-Jewish Refugee in Vichy France, 1939–1941: Arno Motulsky's Memoir of Life in the Internment Camps at St. Cyprien and Gurs." *American Journal of Medical Genetics* 176A:1289–95.

Munroe, J. Brian. 2004. "A Coalition to Drive Personalized Medicine Forward." *Personalized Medicine* 1:9–13.

Naik, Guatam. 2013. "A Genetic Code for Genius?" *Wall Street Journal,* Feb. 15, 2013. www.wsj.com.

NASA. 2004. "President Bush Offers New Vision for NASA." Jan. 14, 2004. www.nasa.gov.

National Children's Study Interagency Coordinating Committee. 2003. "The National Children's Study of Environmental Effects on Child Health and Development." *Environmental Health Perspectives* 111:642–46.

National Children's Study Working Group. 2014. "Final Report." acd.od.nih.gov.

National Human Genome Research Institute. 2000. "Remarks Made by the President, Prime Minister Tony Blair of England (via Satellite), Dr. Francis Collins, Director of the National Human Genome Research Institute, and Dr. Craig Venter, President and Chief Scientific Officer, Celera Genomics Corporation, on the Completion of the First Survey of the Entire Human Genome Project." June 26, 2000. www.genome.gov.

———. 2003. "International Consortium Completes Human Genome Project: All Goals Achieved; New Vision for Genome Research Unveiled." www.genome.gov.

Nebert, Daniel W., Milton Adesnik, Minor J. Coon, et al. 1987. "The P450 Gene Superfamily: Recommended Nomenclature." *DNA* 6:1–11.

Nebert, Daniel W., and Frank J. Gonzalez. 1987. "P450 Genes: Structure, Evolution, and Regulation." *Annual Review of Biochemistry* 56:945–93.

Nebert, Daniel W., Lucia Jorge-Nebert, and Elliot S. Vesell. 2003. "Pharmacogenomics and 'Individualized Drug Therapy': High Expectations and Disappointing Achievements." *American Journal of Pharmacogenomics* 3:361–70.

Nebert, Daniel W., and David W. Russell. 2002. "Clinical Importance of the Cytochromes P450." *Lancet* 360:1155–62.

Nebert, Daniel W., Ge Zhang, and Elliot S. Vesell. 2008. "From Human Genetics and Genomics to Pharmacogenetics and Pharmacogenomics: Past Lessons, Future Directions." *Drug Metabolism Review* 40:187–224.

Neel, James V. 1962. "Diabetes Mellitus: A 'Thrifty' Genotype Rendered Detrimental by 'Progress'?" *American Journal of Human Genetics* 14:353–62.

———. 1999. "The 'Thrifty Genotype' in 1998." *Nutrition Reviews* 57:S2–S9.

Neel, James V., Alan B. Weder, and Stevo Julius. 1998. "Type II Diabetes, Essential Hypertension, and Obesity as 'Syndromes of Impaired Genetic Homeostasis': The 'Thrifty Genotype' Hypothesis Enters the 21st Century." *Perspectives in Biology and Medicine* 42:44–74.

Neergaard, Lauren. 2017. "Wanted: 1 Million People to Study Genes, Habits, Health." Associated Press, Sept. 25, 2017. apnews.com.

Nelson, Robert G., Peter H. Bennett, Gerald J. Beck, et al. 1996. "Development and Progression of Renal Disease in Pima Indians with Non-Insulin-Dependent Diabetes Mellitus." *New England Journal of Medicine* 335:1636–42.

Nelson, Roxanne. 2019. "PriceMedicine." Last for Transformative Cancer Drug." Medscape, Dec. 19, 2019. www.medscape.com.

Newman, Richard S. 2016. *Love Canal: A Toxic History from Colonial Times to the Present.* New York: Oxford University Press.

New York Times. 1962. "Medicine and Genetics." Oct. 13, 1962, 24.

———. 1973. "High Levels of Lead in Texas Linked to El Paso Smelter." Dec. 18, 1973, 27.

———. 1994. "U.S. Policy: 'Back to Sleep' for Babies." June 22, 1994.

———. 1996. "'Back to Sleep' Effort Is Said to Save 1,500." June 26, 1996. www.nytimes.com.

NIH. 2006. "Two NIH Initiatives Launch Intensive Effort to Determine Genetic and Environmental Roots of Common Diseases." News release, Feb. 8, 2006. www.nih.gov.

———. 2018. "NIH Announces National Enrollment Date for All of Us Research Program to Advance Precision Medicine." News release, May 1, 2018. www.nih.gov.

NIH Record. 1966. "New Portable Clinical Facility Dedicated in Arizona, Used for Long-Range Studies." 43:1, 7.

Nowell, Peter. 2007. "Discovery of the Philadelphia Chromosome: A Personal Perspective." *Journal of Clinical Investigation* 117:2033–35.

Oakes, Jeannie. 1982. "The Reproduction of Inequality: The Content of Secondary School Tracking." *Urban Review* 14:107–20.

———. 1985. *Keeping Track: How Schools Structure Inequality.* New Haven, Conn.: Yale University Press.

Oestreicher, Paul. 2002. "Genaissance Pharmaceuticals, Inc." *Pharmacogenomics* 3:273–76.

Office of Management Assessment. 2009. "Internal Control Review of the National Children's Study." July 10, 2009.

Offord, Catherine. 2018. "Father of Pharmacogenetics Dies." *Scientist*, Jan. 31, 2018. www.the-scientist.com.

Ogilvy Public Relations Worldwide. 2004. "The Early Planning and Development of the National Children's Study: A Historical Perspective."

———. 2007. "The Early Implementation of the National Children's Study: A Historical Perspective Updated."

Okie, Susan. 2001. "Cancer Drug Approved Quickly." *Washington Post*, May 11, 2001. www.washingtonpost.com.

Olby, Robert. 1974. *The Path to the Double Helix: The Discovery of DNA*. Seattle: The University of Washington Press.

———. 1985. *The Origins of Mendelism*. Chicago: University of Chicago Press.

Ollier, William, Tim Sprosen, and Tim Peakman. 2005. "UK Biobank: From Concept to Reality." *Pharmacogenomics* 6:639–46.

Olson, Maynard. 2017. "A Behind-the-Scenes Story of Precision Medicine." *Genomics Proteomics Bioinformatics* 15:3–10.

Oreskes, Naomi, and Erik Conway. 2010. *Merchants of Doubt: How a Handful of Scientists Obscured the Truth on Issues from Tobacco Smoke to Climate Change*. London: Bloomsbury Press.

Osler, William. 1892. *Principles and Practice of Medicine*. New York: D. Appleton.

Ozanne, S. E., and C. N. Hales. 1998. "Thrifty Yes, Genetic No." *Diabetologia* 41:485–87.

Page, Kathleen A., Shan Luo, Xinhui Wang, et al. 2019. "Children Exposed to Maternal Obesity or Gestational Diabetes Mellitus During Early Fetal Development Have Hypothalamic Alterations That Predict Future Weight Gain." *Diabetes Care* 42:1473–80.

Panel to Review the National Children's Study Research Plan. 2008. *The National Children's Study Research Plan: A Review*. Washington, D.C.: National Academies Press.

Panofsky, Aaron. 2014. *Misbehaving Science: Controversy and the Development of Behavior Genetics*. Chicago: University of Chicago Press.

———. 2015. "What Does Behavioral Genetics Offer for Improving Education?" *Hastings Center Report*, 45:S43–S49.

Paradies, Yin C., Michael J. Montoya, and Stephanie M. Fullerton. 2007. "Racialized Genetics and the Study of Complex Diseases: The Thrifty Genotype Revisited." *Perspectives in Biology and Medicine* 50:203–27.

Parens, Erik. 2020. "The Inflated Promise of Genomic Medicine." *Scientific American*, June 1, 2020. blogs.scientificamerican.com.

Parsons, Deirdre Bradford. 2007. "Seminal Genomic Technologies: Illumina, Inc. & High-Throughput SNP Genotyping BeadArray Technology, a Case Study." Master's thesis, Duke University.

Patel, Chirag J., Jayanta Bhattacharya, and Atul J. Butte. 2010. "An Environment-Wide Association Study (EWAS) on Type 2 Diabetes Mellitus." *PLoS One* 20:e10746.

Paul, Diane B. 1995. *Controlling Human Heredity: 1865 to the Present.* Atlantic Highlands, N.J.: Humanities Press.

Pauling, Linus, Harvey A. Itano, S. J. Singer, et al. 1949. "Sickle Cell Anemia, a Molecular Disease." *Science* 110:543–48.

Payne-Sturges, Devon C., Gilbert C. Gee, and Deborah A. Cory-Slechta. 2021. "Confronting Racism in Environmental Health Science: Moving the Science Forward for Eliminating Racial Inequities." *Environmental Health Perspectives* 129:055002–1.

PBS. 1998. *Faith and Reason.* Sept. 11, 1998.

Pearson, Helen. 2003. "Artificial Nose Apes Dog Snout." *Nature,* April 7, 2003. www.nature.com.

Peck, Peggy. 2001. "New Leukemia Drug Works for Incurable Stomach Tumors, Too." WebMD, May 13, 2001. www.webmd.com.

Pennisi, Elizabeth. 2000. "Finally, the Book of Life and Instructions for Navigating It." *Science* 288:2304–7.

Perales, Monica. 2008. "Fighting to Stay in Smeltertown: Lead Contamination and Environmental Justice in a Mexican American Community." *Western Historical Quarterly* 39:41–63.

———. 2010. *Smeltertown: Making and Remembering a Southwest Border Community.* Chapel Hill: University of North Carolina Press.

Perlman, Robert L., and Diddahally R. Govindaraju. 2016. "Archibald E. Garrod: The Father of Precision Medicine." *Genetics in Medicine* 18:1088–89.

Persaud, Krishna, and George Dodd. 1982. "Analysis of Discrimination Mechanisms in the Mammalian Olfactory System Using a Model Nose." *Nature* 299:352–55.

Personalized Medicine Coalition. 2017. *The Personalized Medicine Report: Opportunities, Challenges, and the Future.* www.personalizedmedicinecoalition.org.

———. 2018. *Personalized Medicine at the FDA: A Progress and Outlook Report.* www.personalizedmedicinecoalition.org.

———. 2020. *The Personalized Medicine Report: Opportunities, Challenges, and the Future.* www.personalizedmedicinecoalition.org.

Pettitt, David J., Robert G. Nelson, Mohammed F. Saad, et al. 1993. "Diabetes and Obesity in the Offspring of Pima Indian Women with Diabetes During Pregnancy." *Diabetes Care* 16:310–14.

Pettitt, David J., Jennifer Talton, Dana Dabalea, et al. 2014. "Prevalence of Diabetes in U.S. Youth in 2009: The SEARCH for Diabetes in Youth Study." *Diabetes Care* 37:402–8.

Phillips, Christopher. 2020. "Precision Medicine and Its Imprecise History." *Harvard Data Science Review* 2(1): 1–10.

———. Forthcoming. *Number Doctors: The Emergence of Biostatistics and the Reformation of Modern Medicine.*

Piotrowska, Monika. 2009. "What Does It Mean to Be 75% Pumpkin? The Units of Comparative Genomics." *Philosophy of Science* 76:838–50.

Plato. 1925. *Plato in Twelve Volumes.* Vol. 9, translated by Harold N. Fowler. Cambridge, Mass.: Harvard University Press.

Plomin, Robert, Joanna K. J. Kennedy, and Ian W. Craig. 2006. "The Quest for Quantitative Trait Loci Associated with Intelligence." *Intelligence* 34:513–26.

Plomin, Robert, and Jenae Neiderhiser. 1991. "Quantitative Genetics, Molecular Genetics, and Intelligence." *Intelligence* 15:369–87.

Plutynski, Anya. 2018. *Explaining Cancer: Finding Order in Disorder.* Oxford: Oxford University Press.

Pokorska-Bocci, Anna, Alison Stewart, Gurdeep S. Sagoo, et al. 2014. "'Personalized Medicine': What's in a Name?" *Personalized Medicine* 11:197–210.

Popper, Helmut H. 2016. "Progression and Metastasis of Lung Cancer." *Cancer and Metastasis Reviews* 35:75–91.

Porter, Dorothy. 2005. *Health, Civilization, and the State: A History of Public Health from Ancient to Modern Times.* London: Routledge.

Prainsack, Barbara. 2014. "Personhood and Solidarity: What Kind of Personalized Medicine Do We Want?" *Personalized Medicine* 11:651–57.

Prasad, Vinay. 2016. "Perspective: The Precision-Oncology Illusion." *Nature* 537:S63.

———. 2020. *Malignant: How Bad Policy and Bad Evidence Harm People with Cancer.* Baltimore: Johns Hopkins University Press.

President's Council of Advisors on Science and Technology. 2008. *Priorities for Personalized Medicine.*

Prince, Ann. 2012. "Silent No More." *Sanctuary: The Journal of the Massachusetts Audubon Society* 50:7–9.

Proctor, Robert N. 2012. "The History of the Discovery of the Cigarette–Lung Cancer Risk: Evidentiary Traditions, Corporate Denial, Global Toll." *Tobacco Control* 21:87–91.

Prüfer, Kay, Fernando Racimo, Nick Patterson, et al. 2014. "The Complete Genome Sequence of a Neanderthal from the Altai Mountains," *Nature* 505:43–49.

Radcliffe, Shawn. 2019. "Prescription Purgatory: $100,000 a Year to Stay Alive." Healthline, April 5, 2019. www.healthline.com.

Radin, Joanna. 2017. "'Digital Natives': How Medical and Indigenous Histories Make for Big Data." *Osiris* 32:43–64.

Rajan, Kaushik Sunder. 2006. *Biocapital: The Constitution of Postgenomic Life.* Durham, N.C.: Duke University Press.

Ramsey, Scott, David Blough, Anne Kirchhoff, et al. 2013. "Washington State Cancer Patients Found to Be at Greater Risk for Bankruptcy Than People Without a Cancer Diagnosis." *Health Affairs* 32. www.healthaffairs.org.

Ramsey, Scott D., Asthaa Bansal, Catherine R. Fedorenko, et al. 2016. "Financial Insolvency as a Risk Factor for Early Mortality Among Patients with Cancer." *Journal of Clinical Oncology* 20:980–86.

Rasmussen, Frederick N. 2008. "Dr. Janet B. Hardy." *Baltimore Sun,* Oct. 31, 2008. www.baltimoresun.com.

Rasmussen, Morten, Yingrui Li, Stinus Lindgreen, et al. 2010. "Ancient Human Genome Sequence of an Extinct Palaeo-Eskimo." *Nature* 463:757–62.

Ratcliff, Chelsea L., Bob Wong, Jakob D. Jensen, et al. 2021. "The Impact of Communicating Uncertainty on Public Responses to Precision Medicine Research." *Annals of Behavioral Medicine* 55:1048–61.

Ravussin, Eric, Mauro E. Valencia, Julian Esparza, et al. 1994. "Effects of a Traditional Lifestyle on Obesity in Pima Indians." *Diabetes Care* 17:1067–74.

Ray, Keisha. 2021. "In the Name of Racial Justice: Why Bioethics Should Care About Environmental Toxins." *Hastings Center Report* 51:23–26.

———. Forthcoming. *Black Health: The Social, Political, and Cultural Determinants of Black People's Health.* New York: Oxford University Press.

Reich, David. 2018. *Who We Are and How We Got Here: Ancient DNA and the New Science of the Human Past.* New York: Pantheon Books.

Reich, David, Richard E. Green, Martin Kircher, et al. 2010. "Genetic History of an Archaic Hominin Group from Denisova Cave in Siberia." *Nature* 468:1053–60.

Reinagel, Monica. 2012. "Can Genetic Testing Reveal Your Ideal Diet?" Quick and Dirty Tips, Sept. 4, 2012. www.quickanddirtytips.com.

Remon, Jordi, Julia Bonastre, and Benjamin Besse. 2016. "The 5000% Case: A Glimpse into the Financial Issue of Lung Cancer Treatment." *European Respiratory Journal* 47:1331–33.

Renfrew, Colin, and Paul Bahn, eds. 2005. *Archaeology: The Key Concepts.* London: Routledge.

Republican Party Platform. 2008. www.presidency.ucsb.edu.

Research!America. 2017. "U.S. Investments in Medical and Health Research and Development, 2013–2016." www.researchamerica.org.

———. 2018. "U.S. Investments in Medical and Health Research and Development, 2013–2017." www.researchamerica.org.

———. 2019. "U.S. Investments in Medical and Health Research and Development, 2013–2018." www.researchamerica.org.

Rich, Andrew. 2004. *Think Tanks, Public Policy, and the Politics of Expertise.* Cambridge, U.K.: Cambridge University Press.

Richardson, Sarah S., and Hallam Stevens, eds. 2015. *Postgenomics: Perspectives on Biology After the Genome.* Durham, N.C.: Duke University Press.

Rietveld, Cornelius A., Sarah E. Medland, Jaime Derringer, et al. 2013. "GWAS of 126,559 Individuals Identifies Genetic Variants Associated with Educational Attainment." *Science* 340:1467–71.

Riordan, John R., Johanna M. Rommens, Bat-sheva Kerem, et al. 1989. "Identification of the Cystic Fibrosis Gene and Characterization of Complementary DNA." *Science* 245:1066–73.

Risch, Neil. 2005. "The SNP Endgame: A Multidisciplinary Approach." *American Journal of Human Genetics* 76:221–26.

Risch, Neil, and Kathleen Merikangas. 1996. "The Future of Genetic Studies of Complex Traits." *Science* 273:1516–17.

Risk Policy Report. 2008. "With OMB Approval, Children's Study Expands Centers, Sets Timeline." 15:7–8.

Rist, Ray C. 1970. "Student Social Class and Teacher Expectations: The Self-Fulfilling Prophecy in Ghetto Education." *Harvard Educational Review* 70:257–301.

Roberts, Dorothy. 2008. "Is Race-Based Medicine Good for Us? African American Approaches to Race, Biomedicine, and Equality." *Journal of Law, Medicine, and Ethics* 36:537–45.

———. 2011a. *Fatal Invention: How Science, Politics, and Big Business Re-create Race in the Twenty-First Century.* New York: New Press.

———. 2011b. "What's Wrong with Race-Based Medicine? Genes, Drugs, and Health Disparities." *Minnesota Journal of Law, Science, and Technology* 12:1–21.

———. 2012. "Debating the Cause of Health Disparities: Implications for Bioethics and Racial Equity." *Cambridge Quarterly of Healthcare Ethics* 21:332–41.

———. 2015. "Can Research on the Genetics of Intelligence Be 'Socially Neutral'?" *Hastings Center Report* 45:S50–S53.

Roberts, J. Timmons, and Melissa M. Toffolon-Weiss. 2001. *Chronicles from the Environmental Justice Frontline.* Cambridge, U.K.: Cambridge University Press.

Roberts, Jessica L. 2010. "Preempting Discrimination: Lessons from the Genetic Information Nondiscrimination Act." *Vanderbilt Law Review* 63:439–90.

———. 2011. "The Genetic Information Nondiscrimination Act as an Antidiscrimination Law." *Notre Dame Law Review* 86:597–648.

Roberts, Robert. 2008. "Personalized Medicine: A Reality Within This Decade." *Journal of Cardiovascular Translational Research* 1:11–16.

Roden, D. M., and R. F. Tyndale. 2013. "Genomic Medicine, Precision Medicine, Personalized Medicine: What's in a Name?" *Clinical Pharmacology and Therapeutics* 94:169–72.

Rodengen, Jeffrey L. 1999. *The Legend of Pfizer.* Fort Lauderdale, Fla.: Write Stuff Syndicate.

Roehr, Bob. 2018. "Arno Motulsky: Father of Pharmacogenetics." *British Medical Journal* 360:k884.

Rogers, Alan R., Ryan J. Bohlender, and Chad D. Huff. 2017. "Early History of Neanderthals and Denisovans." *Proceedings of the National Academy of Sciences* 114:9859–63.

Romano, Lois. 1983. "Rita Lavelle, Dumped." *Washington Post,* March 5, 1983. www.washingtonpost.com.

Romano, Lois, and Jacqueline Trescott. 1983. "The Rise and Fall of Anne Burford." *Washington Post,* March 10, 1983. www.washingtonpost.com.

Romero, Mary. 1984. "The Death of Smeltertown: A Case Study of Lead Poisoning in a Chicano Community." In *The Chicano Struggle: Analysis of Past and Present Efforts,* edited by John Garcia, Juan Garcia, and Teresa Cordova, 26–41. Tempe, Ariz.: Bilingual Press.

Ropiek, Annie. 2018. "Military to Study Past Chemical Use at Former Pease Air Force Base." *Air Force Times,* Dec. 10, 2018. www.airforcetimes.com.

Rosen, Christine. 2004. *Preaching Eugenics: Religious Leaders and the American Eugenics Movement.* New York: Oxford University Press.

Rosen, George. (1958) 2015. *A History of Public Health.* Baltimore: Johns Hopkins University Press.

Rosenbaum, James E. 1976. *Making Inequality: The Hidden Curriculum of High School Tracking.* New York: John Wiley & Sons.

Rosner, David, and Gerald Markowitz. 2005. "Standing Up to the Lead Industry: An Interview with Herbert Needleman." *Public Health Reports* 120:330–37.

Rosner, Fred. 1969. "Hemophilia in the Talmud and Rabbinic Writings." *Annals of Internal Medicine* 70:833–37.

Rothstein, Mark A. 2008. "Putting the Genetic Information Nondiscrimination Act in Context." *Genetics in Medicine* 10:655–56.

———. 2017. "Structural Challenges of Precision Medicine." *Journal of Law, Medicine, and Ethics* 45:274–79.

Rowley, Erin. 2016. "Genomic Testing: An Approach to Treatment That Is Promising, but Not Yet Standard." *Insights on Metastatic Breast Cancer* (Fall): 2–3.

Ruaño, Gualberto, Amos S. Deinard, Sarah Tishkoff, et al. 1994. "Detection of DNA Sequence Variation via Deliberate Heteroduplex Formation from Genomic DNAs Amplified En Masse in 'Population Tubes.'" *Genome Research* 3:225–31.

Ruaño, Gualberto, and Kenneth K. Kidd. 1992. "Modeling of Heteroduplex Formation During PCR from Mixtures of DNA Templates." *Genome Research* 2:112–16.

Ruiz, Michelle. 2016. "Repealing the Affordable Care Act Could Kill Me." *Vogue*, Dec. 9, 2016. www.vogue.com.

Sabatello, Maya. 2018. "A Genomically Informed Education System? Challenges for Behavioral Genetics." *Journal of Law, Medicine, and Ethics* 46:130–44.

Sabatello, Maya, and Paul S. Appelbaum. 2017. "The Precision Medicine Nation." *Hastings Center Report* 47:19–29.

Sabatello, Maya, Beverly J. Insel, Thomas Corbeil, et al. 2021. "The Double Helix at School: Behavioral Genetics, Disability, and Precision Education." *Social Science and Medicine* 278:113924.

Sahota, Puneet Chawla. 2012. "Genetic Histories: Native Americans' Accounts of Being at Risk for Diabetes." *Social Studies of Science* 42:821–42.

Sankar, Pamela L., and Lisa S. Parker. 2017. "The Precision Medicine Initiative's All of Us Research Program: An Agenda for Research on Its Ethical, Legal, and Social Issues." *Genetics in Medicine* 19:743–50.

Sarche, Michelle, and Paul Spicer. 2008. "Poverty and Health Disparities for American Indian and Alaska Native Children: Current Knowledge and Future Prospects." *Annals of the New York Academy of Sciences* 1136:126–36.

Savica, Rodolfo, Brandon R. Grossardt, James H. Bower, et al. 2016. "Time Trends in the Incidence of Parkinson's Disease." *JAMA Neurology* 73:981–89.

Savoie, Hillary. 2015a. *Around and into the Unknown.* Glasgow: Ponies + Horses Books.

———. 2015b. "When Your Child Isn't Just Rare but Probably One of a Kind." *Complex Child*, May 18, 2015. complexchild.org/articles.

———. 2019. "On Parenting from the Place Where Science, Medicine, and Love Collide." *Narrative Inquiry in Bioethics* 9:8–11.

Sawchuk, Elizabeth, and Mary Prendergast. 2019. "Ancient DNA Is a Powerful Tool for Studying the Past—When Archaeologists and Geneticists Work Together." *Conversation*, March 11, 2019. theconversation.com.

Schaffner, Kenneth F. 2016. *Behaving: What's Genetic, What's Not, and Why Should We Care?* Oxford: Oxford University Press.

Schilsky, Richard L. 2020. "Talking the Talk About Tumor Genomic Testing." *Journal of the National Cancer Institute* 112:436–37.

Schmeck, Harold M., Jr. 1962. "Heredity Linked to Drug Effects." *New York Times*, Oct. 10, 1962, 96.

Schmidt, Charles. 2016. "The Death of the National Children's Study: What Went Wrong?" *Undark*, May 25, 2016. undark.org.

Schmidt, Karen F. 1998. "Just for You." *New Scientist* 160:32–36.

Schneider, Susanne A., and Roy N. Alcalay. 2020. "Precision Medicine in Parkinson's Disease: Emerging Treatments for Genetic Parkinson's Disease." *Journal of Neurology* 267:860–69.

Schoen, Johanna. 2005. *Choice and Coercion: Birth Control, Sterilization, and Abortion in Public Health and Welfare.* Chapel Hill: University of North Carolina Press.

Schork, Nicholas J., Sarah S. Murray, Kelly A. Frazer, et al. 2009. "Common vs. Rare Allele Hypotheses for Complex Diseases." *Current Opinion in Genetics and Development* 19:212–19.

Schudel, Matt. 2008. "Janet Hardy; Pioneering Pediatrics Professor." *Washington Post,* Nov. 9, 2008. www.washingtonpost.com.

Schulman, Bruce J., and Julian E. Zelizer, eds. 2008. *Rightward Bound: Making America Conservative in the 1970s.* Cambridge, Mass.: Harvard University Press.

Schulz, Leslie O., and Lisa S. Chaudhari. 2015. "High-Risk Populations: The Pimas of Arizona and Mexico." *Current Obesity Reports* 4:92–98.

Scientific American. 2006. "Don't Rob the Cradle." May 1, 2006. www.scientificamerican.com.

Scott, Robert. 1970a. "Health Care Priority and Sickle Cell Anemia." *Journal of the American Medical Association* 214:731–34.

———. 1970b. "Sickle Cell Anemia: High Prevalence and Low Priority." *New England Journal of Medicine* 282:164–65.

Scudelleri, Megan. 2012. "Upheaval over Children's Health Study." *Scientist,* March 21, 2012. www.the-scientist.com.

Secretary's Advisory Committee on Genetics, Health, and Society. 2008. *Realizing the Potential of Pharmacogenomics: Opportunities and Challenges.*

Sellers, Christopher. 2012. *Crabgrass Crucible: Suburban Nature and the Rise of Environmentalism in the Twentieth Century.* Chapel Hill: University of North Carolina Press.

Shabecoff, Philip. 1982. "House Charges Head of E.P.A. with Contempt." *New York Times,* Dec. 17, 1982. www.nytimes.com.

Shastry, B. S. 2006. "Pharmacogenomics and the Concept of Individualized Medicine." *Pharmacogenomics Journal* 6:16–21.

Shreeve, James. 2004. *The Genome War: How Craig Venter Tried to Capture the Code of Life and Save the World.* New York: Alfred A. Knopf.

Sideris, Lisa H., and Kathleen Dean Moore, eds. 2007. *Rachel Carson: Legacy and Challenge.* Albany: State University of New York Press.

Sipher, Devan. 2015. "Unbreakable Bonds in the Folds of a Wedding Dress." *New York Times,* May 8, 2015. www.nytimes.com.

Skiotis, Gerasimos P., George D. Kalliolias, and Athanasios G. Papavassiliou. 2005. "Pharmacogenetic Principles in the Hippocratic Writings." *Journal of Clinical Pharmacology* 45:1218–20.

Skloot, Rebecca. 2010. *The Immortal Life of Henrietta Lacks.* New York: Crown.

Smith, Lauren A., Suzette O. Oyeku, Charles Homer, et al. 2006. "Sickle Cell Disease: A Question of Equity and Quality." *Pediatrics* 117:1763–70.

Smith, Michael B. 2001. "'Silence, Ms. Carson!': Science, Gender, and the Reception of *Silent Spring*." *Feminist Studies* 27:733–52.

Smith-Morris, Carolyn. 2006. *Diabetes Among the Pima: Stories of Survival*. Tucson: University of Arizona Press.

Snow, Stephanie J. 2008. *Blessed Days of Anaesthesia: Anaesthetics Changed the World*. Oxford: Oxford University Press.

Snyder, Alison. 2014. "Obituary: Robert E. Cooke." *Lancet* 383:1378.

Solomon, Miriam. 2016. "On Ways of Knowing in Medicine." *CMAJ* 188:289–90.

SoRelle, Jeffrey A., Drew M. Thodeson, Susan Arnold, et al. 2019. "Clinical Utility of Reinterpreting Previously Reported Genomic Epilepsy Test Results for Pediatric Patients." *JAMA Pediatrics* 173:e182302.

Souder, William. 2012. *On a Farther Shore: The Life and Legacy of Rachel Carson*. New York: Crown.

Speake, Georgina, Brian Holloway, and Gerard Costello. 2005. "Recent Developments Related to the EGFR as a Target for Cancer Chemotherapy." *Current Opinion in Pharmacology* 5:343–49.

Specter, Michael. 2014. "The Gene Factory." *New Yorker*, Jan. 6, 2014. www.newyorker.com.

Spezio, Teresa Sabol. 2018. *Slick Policy: Environmental and Science Policy in the Aftermath of the Santa Barbara Oil Spill*. Pittsburgh: University of Pittsburgh Press.

Stahl, Jason. 2016. *Right Moves: The Conservative Think Tank in the American Political Culture Since 1945*. Chapel Hill: University of North Carolina Press.

Starr, Douglas. 1998. *Blood: An Epic History of Medicine and Commerce*. New York: Alfred A. Knopf.

Stegenga, Jacob. 2018. *Care and Cure: An Introduction to Philosophy of Medicine*. Chicago: University of Chicago Press.

Stegman, Stephanie. 2010. "Taking Control: Fifty Years of Diabetes in the American Southwest, 1940–1990." PhD diss., Arizona State University.

Stern, Alexandra Minna. 2005. *Eugenic Nation: Faults and Frontiers of Better Breeding in Modern America*. Berkeley: University of California Press.

———. 2012. *Telling Genes: The Story of Genetic Counseling in America*. Baltimore: Johns Hopkins University Press.

Stipp, David. 2000. "13 Biotech IPOs to Watch For." *Fortune*, April 3, 2000. archive.fortune.com.

Stitzel, Shannon E., Deborah R. Stein, and David R. Walt. 2003. "Enhancing Vapor Sensor Discrimination by Mimicking a Canine Nasal Cavity Flow Environment." *Journal of the American Chemical Society* 125:3684–85.

Stix, Gary. 1998. "Personal Pills." *Scientific American*, Oct. 1998, 17–18.

Straatsma, Bradley R. 2018. "Precision Medicine and Clinical Ophthalmology." *Indian Journal of Ophthalmology* 66:1389–90.

Stradling, David, and Richard Stradling. 2008. "Perceptions of the Burning River: Deindustrialization and Cleveland's Cuyahoga River." *Environmental History* 13:515–35.

Strouse, John J., Katie Lobner, Sophie Lanzkron, et al. 2013. "NIH and National Foundation Expenditures for Sickle Cell Disease and Cystic Fibrosis Are Associated with Pubmed Publications and FDA Approvals." *Blood* 122:1739.

Struck, Kathleen. 2013. "Big Changes for U.S. Child Study." *MedPage Today,* Feb. 4, 2013. www.medpagetoday.com.

Sullivan, Marianne. 2014. *Tainted Earth: Smelters, Public Health, and the Environment.* New Brunswick, N.J.: Rutgers University Press.

Sulston, John, and Georgina Ferry. 2002. *The Common Thread: A Story of Science, Politics, Ethics, and the Human Genome.* Washington, D.C.: Joseph Henry Press.

Swain, Rajesh. 2006. "Genomics, Personalized Medicine Act Introduced by Sen. Obama." *US Fed News,* Aug. 11, 2006.

Szabo, Liz. 2017. "Widespread Hype Gives False Hope to Many Cancer Patients." Kaiser Health News, April 27, 2017. khn.org.

———. 2018a. "Much Touted for Cancer, 'Precision Medicine' Often Misses the Target." Kaiser Health News, Sept. 13, 2018. khn.org.

———. 2018b. "Pricey Precision Medicine Often Financially Toxic for Cancer Patients." Kaiser Health News, Nov. 1, 2018. khn.org.

Tabb, Kathryn, and Maël Lemoine. 2021. "The Prospects of Precision Psychiatry." *Theoretical Medicine and Bioethics* 42:193–210.

Tabery, James. 2014. *Beyond Versus: The Struggle to Understand the Interaction of Nature and Nurture.* Cambridge, Mass.: MIT Press.

Tartarone, Alfredo, Chiara Lazzari, Rosa Lerose, et al. 2013. "Mechanisms of Resistance to EGFR Tyrosine Kinase Inhibitors Gefitinib/Erlotinib and to ALK Inhibitor Crizotinib." *Lung Cancer* 81:328–36.

Task Force on Sudden Infant Death Syndrome. 2005. "The Changing Concept of Sudden Infant Death Syndrome: Diagnostic Coding Shifts, Controversies Regarding the Sleeping Environment, and New Variables to Consider in Reducing Risk." *Pediatrics* 116:1245–55.

Taylor, Dorceta E. 2014. *Toxic Communities: Environmental Racism, Industrial Pollution, and Residential Mobility.* New York: New York University Press.

Taylor, Lynne. 2008. "Bush Health Chief Urges Obama: 'Focus on Personalised Health Care.'" PharmaTimes, Nov. 17, 2008. www.pharmatimes.com.

Technology Networks. 2006. "U.S. Senator Introduces Bill to Help Tap Power of Genomics to Find Cures." Aug. 11, 2006. www.technologynetworks.com.

Tessema, Frazer A., Aaron S. Kesselheim, and Michael S. Sinha. 2020. "Generic but Expensive: Why Prices Can Remain High for Off-Patent Drugs." *Hastings Law Journal* 71:1019–52.

Thacker, Stephen B., Andrew L. Dannenberg, and Douglas H. Hamilton. 2001. "Epidemic Intelligence Service of the Centers for Disease Control and Prevention: 50 Years of Training and Service in Applied Epidemiology." *American Journal of Epidemiology* 154:985–92.

Thompson, Dick. 1999. "Craig Venter. Gene Maverick." *Time,* Jan. 11, 1999. content.time.com.

Thorisson, Gudmundur A., Albert V. Smith, Lalitha Krishnan, et al. 2005. "The International HapMap Project Website." *Genome Research* 15:1592–93.

Thorsby, E. 2009. "A Short History of HLA." *Tissue Antigens* 74:101–16.

Timmerman, Luke. 2013. "What's in a Name? A Lot When It Comes to 'Precision Medicine.'" Xconomy, Feb. 4, 2013. xconomy.com.

Tozzi, John, and Alex Wayne. 2014. "How the U.S. Government Botched Its Multi-Billion Dollar Plan to Beat Childhood Disease." Bloomberg, Dec 23, 2014. bloomberg.com.

Trigger, Bruce G. 1989. *A History of Archaeological Thought.* Cambridge, U.K.: Cambridge University Press.

Trusheim, Mark R., Ernst R. Berndt, and Frank L. Douglas. 2007. "Stratified Medicine: Strategic and Economic Implications of Combining Drugs and Clinical Biomarkers." *Nature Reviews Drug Discovery* 6:287–93.

Tsao, Connie W., and Ramachandran S. Vasan. 2015. "Cohort Profile—the Framingham Heart Study (FHS): Overview of Milestones in Cardiovascular Epidemiology." *International Journal of Epidemiology* 44:1800–1833.

Tsosie, Krystal S., Joseph M. Yracheta, and Donna Dickenson. 2019. "Overvaluing Individual Consent Ignores Risks to Tribal Participants." *Nature Reviews Genetics* 20: 497–98.

Tsosie, Krystal S., Rene L. Begay, Keolu Fox, et al. 2020. "Generations of Genomes: Advances in Paleogenomics Technology and Engagement for Indigenous People of the Americas." *Current Opinion in Genetics & Development* 62:91–96.

Tsosie, Krystal S., Joseph M. Yracheta, Jessica A. Kolopenuk, et al. 2021. "We Have 'Gifted' Enough: Indigenous Genomic Data Sovereignty in Precision Medicine." *The American Journal of Bioethics* 21:72–75.

Tsosie, Krystal S., Keolu Fox, and Joseph M. Yracheta. 2021. "Genomics Data: The Broken Promise Is to Indigenous People." *Nature* 591:529.

Tsui, Lap-Chee, and Ruslan Dorfman. 2013. "The Cystic Fibrosis Gene: A Molecular Genetic Perspective." *Cold Spring Harbor Perspectives in Medicine* 3:a009472.

Tuchman, Arleen Marcia. 2020. *Diabetes: A History of Race and Disease.* New Haven, Conn.: Yale University Press.

Turkheimer, Eric. 1998. "Heritability and Biological Explanation." *Psychological Review* 105:782–91.

———. 2000. "Three Laws of Behavioral Genetics and What They Mean." *Current Directions in Psychological Science* 9:160–64.

———. 2011. "Still Missing." *Research in Human Development* 8:227–41.

———. 2014. "Genome Wide Association Studies Are Social Science." In *Philosophy of Behavioral Biology,* edited by Kathryn Plaisance and Thomas Reydon, 43–64. Dordrecht: Springer.

———. 2015. "Genetic Prediction." *Hastings Center Report* 45:S32–S38.

———. 2016. "Weak Genetic Explanation 20 Years Later: Reply to Plomin et al. (2016)." *Perspectives on Psychological Science* 11:24–28.

———. 2019a. "The Social Science Blues." *Hastings Center Report* 49:45–47.

———. 2019b. "The Shiny—and Potentially Dangerous—New Tool for Predicting Human Behavior." Leaps.org, Aug. 22, 2019. leaps.org.

Turner, James Morton, and Andrew C. Isenberg. 2018. *The Republican Reversal: Conservatives and the Environment from Nixon to Trump.* Cambridge, Mass.: Harvard University Press.

Tutton, Richard. 2012. "Personalizing Medicine: Futures Present and Past." *Social Science and Medicine* 75:1721–28.

Tutton, Richard, and Kimberly Jamie. 2013. "Personalized Medicine in Context: Social Science Perspectives." *Drug Discovery Today: Therapeutic Strategies* 10:e183–e187.

U.S. Department of Health and Human Services. 2007. *Personalized Health Care: Opportunities, Pathways, Resources.*

———. 2008. *Personalized Health Care: Pioneers, Partnerships, Progress.*

U.S. General Accounting Office. 1983. "Siting of Hazardous Waste Landfills and Their Correlation with Racial and Economic Status of Surrounding Communities." www.gao.gov.

UT Southwestern Medical Center. 2018. "Reanalyzing Gene Tests Prompts New Diagnoses in Kids." Press release, Nov. 5, 2018. www.utsouthwestern.edu.

Valdes, Roland, Jr., and DeLu (Tyler) Yin. 2016. "Fundamentals of Pharmacogenetics in Personalized, Precision Medicine." *Clinical and Laboratory Medicine* 36:447–59.

Valles, Sean A. 2012. "Should Direct-to-Consumer Personalized Genomic Medicine Remain Unregulated? A Rebuttal of the Defenses," *Perspectives in Biology and Medicine* 55: 250–65.

———. 2018. *Philosophy of Population Health: Philosophy for a New Public Health Era.* London: Routledge.

———. 2020. "Philosophy of Biomedicine." In *The Stanford Encyclopedia of Philosophy*, edited by Edward N. Zalta. plato.stanford.edu.

Van Arnum, Patricia. 2012. "Decades of Change for the Top Pharmaceutical Companies." PharmaTech.com, July 11, 2012. www.pharmtech.com.

Vasella, Daniel. 2003. *Magic Cancer Bullet: How a Tiny Orange Pill Is Rewriting Medical History.* With Robert Slater. New York: HarperBusiness.

Vassey, Jason L., Marie-France Hivert, Bianca Porneala, et al. 2014. "Polygenic Type 2 Diabetes Prediction at the Limit of Common Variant Detection." *Diabetes* 63:2172–82.

Veeramah, Krishna. 2018. "The Importance of Fine-Scale Studies for Integrating Paleogenomics and Archaeology." *Current Opinion in Genetics and Development* 53:83–89.

Villagran, Lauren. 2016. "Before Flint, Before East Chicago, There Was Smeltertown." National Resources Defense Council, Nov. 29, 2016. www.nrdc.org.

Visscher, Peter M., Matthew A. Brown, Mark I. McCarthy, et al. 2012. "Five Years of GWAS Discovery." *American Journal of Human Genetics* 90:7–24.

Vogel, Friedrich. 1959. "Moderne Probleme der Humangenetik." *Ergebnisse der Inneren Medizin und Kinderheilkunde* 12:52–125.

Vogt, Henrik, Sara Green, and John Brodersen. 2018. "Precision Medicine in the Clouds." *Nature Biotechnology* 36:678–80.

von Linstow, Christian U., Ziv Gan-Or, and Patrik Brundin. 2020. "Precision Medicine in Parkinson's Disease Patients with *LRRK2* and *GBA* Risk Variants—Let's Get Even More Personal." *Translational Neurodegeneration* 9:39.

Waalen, Jill. 2001. "Gleevec's Glory Days." *HHMI Bulletin*, Dec. 2001, 10–15.

Wade, Nicholas. 1999. "The Genome's Combative Entrepreneur." *New York Times*, May 18, 1999. www.nytimes.com.

———. 2001. "Swift Approval for a New Kind of Cancer Drug." *New York Times*, May 11, 2001. www.nytimes.com.

———. 2004. "Human Gene Total Falls Below 25,000." *New York Times*, Oct. 21, 2004.

Wadman, Meredith. 2008. "NIH Director Resigns." *Nature*, Sept. 24, 2008. www.nature.com.

———. 2009. "Congress Probes NIH Stimulus Funds." *Nature* 458:556.

———. 2012a. "Growing Pains for Children's Study." *Nature* 482:448.

———. 2012b. "Advisor to National Children's Study Quits." *Nature*, March 6, 2012. blogs.nature.com.

———. 2012c. "Child-Study Turmoil Leaves Bitter Taste." *Nature* 485:287–88.

Wailoo, Keith. 2001. *Dying in the City of the Blues: Sickle Cell Anemia and the Politics of Race and Health.* Chapel Hill: University of North Carolina Press.

———. 2017. "Sickle Cell Disease—a History of Progress and Peril." *New England Journal of Medicine* 376:805–7.

Wailoo, Keith, and Stephen Pemberton. 2006. *The Troubled Dream of Genetic Medicine: Ethnicity and Innovation in Tay-Sachs, Cystic Fibrosis, and Sickle Cell Disease.* Baltimore: Johns Hopkins University Press.

Waldman, Scott A., and Andrew Terzic. 2008. "The Roadmap to Personalized Medicine." *Clinical and Translational Science* 1:93.

Walker, Rebekah J., Joni Strom Williams, and Leonard E. Egede. 2016. "Influence of Race, Ethnicity, and Social Determinants of Health on Diabetes Outcomes." *American Journal of Medical Sciences* 351:366–73.

Wallace, M. R., D. A. Marchuk, L. B. Andersen, et al. 1990. "Type 1 Neurofibromatosis Gene: Identification of a Large Transcript Disrupted in Three NF1 Patients." *Science* 249:181–86.

Walt, Christopher Paul. 2005. "Complementing the Genome with an 'Exposome': The Outstanding Challenge of Environmental Exposure Measurement in Molecular Epidemiology." *Cancer Epidemiology, Biomarkers and Prevention* 14:1847–50.

Walt, David R. 2005. "Electronic Noses: Wake Up and Smell the Coffee." *Analytical Chemistry* 77:3.

Walt, David R., Shannon E. Stitzel, and Matthew J. Aernecke. 2012. "Artificial Noses." *American Scientist* 100:38–45.

Wapner, Jessica. 2013. *The Philadelphia Chromosome: A Genetic Mystery, a Lethal Cancer, and the Improbable Invention of a Life-Saving Treatment.* New York: Experiment.

Washburn, Jennifer. 2005. *University, Inc.* New York: Basic Books.

Washington, Harriet A. 2006. *Medical Apartheid: The Dark History of Medical Experimentation on Black Americans from Colonial Times to the Present.* New York: Anchor Books.

———. 2019. *A Terrible Thing to Waste: Environmental Racism and Its Assault on the American Mind.* New York: Little, Brown, Spark.

Washington Post. 2002. "Prime Time's Cancer Cure." Jan. 29, 2002. www.washingtonpost.com.

Watson, James D., and Francis H. C. Crick. 1953. "Molecular Structure of Nucleic Acids: A Structure for Deoxyribose Nucleic Acid." *Nature* 171:737–38.

Weber, Wendell W. 2001. "The Legacy of Pharmacogenetics and Potential Applications." *Mutation Research* 479:1–18.

Webster, Andrew, Paul Martin, Graham Lewis, et al. 2004. "Integrating Pharmacogenetics into Society: In Search of a Model." *Nature Reviews Genetics* 5:663–69.

Wechsler, Jill. 2018. "Precision Medicines Speed Development and Reduce Healthcare Costs." PharmaTech, May 11, 2018. www.pharmtech.com.

Wellcome Trust Press Office. 1999. "SNP Consortium Announced." BioProcess Online, April 19, 1999. www.bioprocessonline.com.

Wells, E. Christian. 2006. *From Hohokam to O'odham: The Protohistoric Occupation of the Middle Gila River Valley, Central Arizona.* Tucson: University of Arizona Press.

Wells, Robert C. 2009. "A New President, a New Congress, and the Path to Personalized Medicine." *Personalized Medicine* 6:235–39.

Wheelock, Anne. 1992. *Crossing the Tracks: How Untracking Can Save America's Schools.* New York: New Press.

White House. 2009. "President Obama Announces Intent to Nominate Francis Collins as NIH Director." Press release, July 8, 2009. obamawhitehouse.archives.gov.

Wilby, Peter. 2014. "Psychologist on a Mission to Give Every Child a Learning Chip." *Guardian,* Feb. 18, 2014. www.theguardian.com/education.

Wilkes, Jennifer J., Gary H. Lyman, David R. Doody, et al. 2019. "Health Care Costs Associated with Contemporary Chronic Myelogenous Leukemia (CML) Therapy Compared to Other Hematologic Malignancies (HEM)." *Blood* 134 (Supplement 1): 4753. doi.org/10.1182/blood-2019-124106.

Willinger, Marian, Howard J. Hoffman, Kuo-Tsung Wu, et al. 1998. "Factors Associated with the Transition to Nonprone Sleep Positions of Infants in the United States: The National Infant Sleep Position Study." *Journal of the American Medical Association* 280:329–35.

Wilson, John P. 2014. *People of the Middle Gila: A Documentary History of the Pimas and Maricopas, 1500s–1945.* Tucson: University of Arizona Press.

Winn, Aaron N., Nancy L. Keating, and Stacie B. Dusetzina. 2016. "Factors Associated with Tyrosine Kinase Inhibitor Initiation and Adherence Among Medicare Beneficiaries with Chronic Myeloid Leukemia." *Journal of Clinical Oncology* 34:4323–28.

Wong, Nathan D., and Daniel Levy. 2013. "Legacy of the Framingham Heart Study: Rationale, Design, Initial Findings, and Implications." *Global Heart* 8:3–9.

Woodson, M. Kyle. 2016. *The Social Organization of Hohokam Irrigation in the Middle Gila River Valley, Arizona.* Tucson: University of Arizona Press.

Woolner, David B. 2017. *The Last 100 Days: FDR at War and at Peace.* New York: Basic Books.

Xie, Hong-Guang, and Felix W. Frueh. 2005. "Pharmacogenomics Steps Towards Personalized Medicine." *Personalized Medicine* 2:325–37.

Yang, Wenya, Jamie L. Hamilton, Catherine Kopil, et al. 2020. "Current and Projected Future Economic Burden of Parkinson's Disease in the U.S." *NPJ Parkinson's Disease* 6:15.

Yong, Ed. 2013. "Chinese Project Probes the Genetics of Genius." *Nature* 497:297–99.

Zafar, S. Yousuf, and Amy P. Abernathy. 2013. "Financial Toxicity, Part I: A New Name for a Growing Problem." *Oncology* 27:80–81.

Zambouri, A. 2007. "Preoperative Evaluation and Preparation for Anesthesia and Surgery." *Hippokratia* 11:13–21.

Zhang, Sarah. 2016. "Illumina, the Google of Genetic Testing, Has Plans for World Domination." *Wired,* Feb. 26, 2016. www.wired.com.

Zhou, Caicun, Yi-Long Wu, Gongyan Chen, et al. 2011. "Erlotinib vs. Chemotherapy as First-Line Treatment for Patients with Advanced EGFR Mutation-Positive Non-Small-Cell Lung Cancer (OPTIMAL, CTONG-0802): A Multicentre, Open-Label, Randomised, Phase 3 Study." *Lancet Oncology* 12:735–42.

Zimring, Carl A. 2015. *Clean and White: A History of Environmental Racism in the United States.* New York: New York University Press.

Index

Page numbers in *italics* refer to illustrations.

A NOTE ABOUT THE AUTHOR

James Tabery is a professor of philosophy at the University of Utah. He is the author of *Beyond Versus: The Struggle to Understand the Interaction of Nature and Nurture.* His work has been featured in *The New York Times, National Geographic,* and *Time* and on National Public Radio. He lives in Salt Lake City with his wife and their three children.

A NOTE ON THE TYPE

This book was set in Minion, a typeface produced by the Adobe Corporation specifically for the Macintosh personal computer, and released in 1990. Designed by Robert Slimbach, Minion combines the classic characteristics of old-style faces with the full complement of weights required for modern typesetting.

COMPOSED BY NORTH MARKET STREET GRAPHICS, LANCASTER, PENNSYLVANIA

PRINTED AND BOUND BY BERRYVILLE GRAPHICS, BERRYVILLE, VIRGINIA

DESIGNED BY MAGGIE HINDERS